Kansas Landscapes

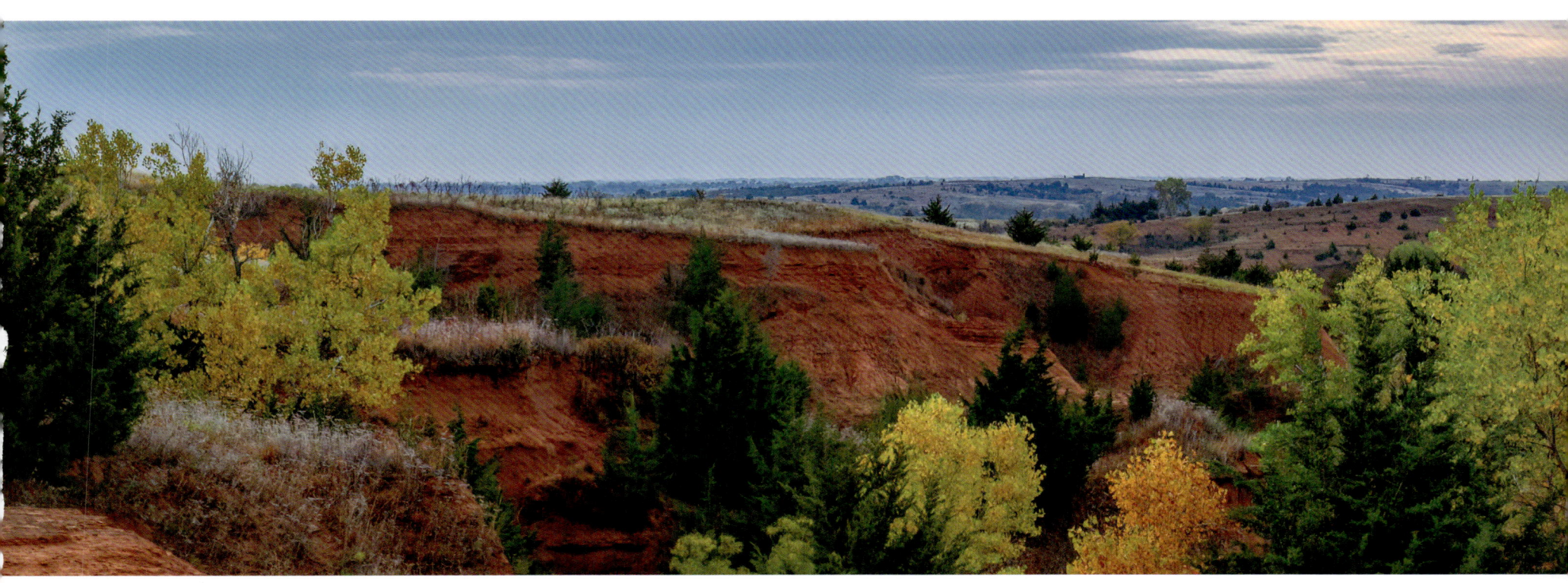

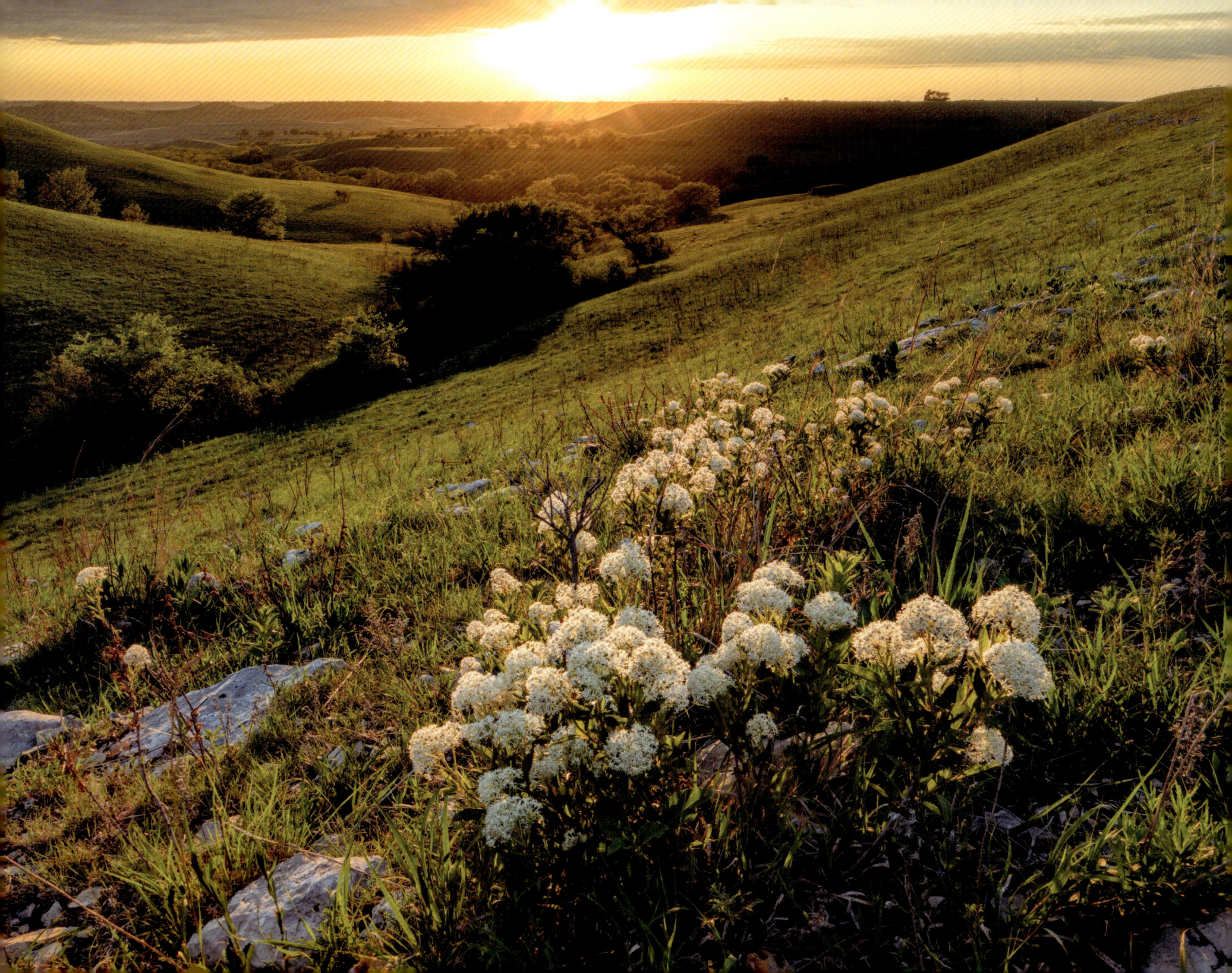

Kansas Landscapes

A Photographic Tour of the Region's Geological History

Keith B. Miller

Photographs by Scott Bean

Additional photographs by
Mark Clarke and Mickey Shannon

University Press of Kansas

All photographs by Scott Bean, unless otherwise credited.
Halftitle Photo: The Red Hills of Barber County
Title Page Photo: The Flint Hills of Pottawatomie County

Published by the University Press of Kansas (Lawrence, Kansas 66045), which was organized by the Kansas Board of Regents and is operated and funded by Emporia State University, Fort Hays State University, Kansas State University, Pittsburg State University, the University of Kansas, and Wichita State University.

Publication made possible, in part, by gifts from:
Sierra Club—Kansas Chapter (PO Box 11415, Overland Park, Kansas 66207-1415, www.sierraclub.org/kansas)
Kansas Geological Foundation (http://www.kgfoundation.org)
West Branch Ridge Runners, Kansas—National Wild Turkey Federation (https://www.nwtf.org/chapters/west-branch-ridge-runners)
Prairie Heritage, Inc. (https://www.prairieheritage.org)
Scott Bean Photography (https://scottbeanphoto.com)
Keith Miller

Interior design by Karl Janssen
Typeset by Blue Heron Typesetters, LLC

Library of Congress Cataloging-in-Publication Data

Names: Miller, Keith B., author
Title: Kansas landscapes : a photographic and geological tour / Keith B. Miller ; photographs by Scott Bean ; additional photographs by Mark Clarke and Mickey Shannon.
Description: Lawrence : University Press of Kansas, 2025.
Identifiers: LCCN 2025008555 | ISBN 9780700638727 (cloth)
Subjects: LCSH: Historical geology—Kansas—Pictorial works | Natural history—Kansas—Pictorial works | Landscape photography—Kansas
Classification: LCC QE113 .M55 2025 | DDC 557.81—dc23/eng/20250324.
LC record available at https://lccn.loc.gov/2025008555.

British Library Cataloguing-in-Publication Data is available.
Authorised Representative Details: Easy Access System Europe
Mustamäe tee 50, 10621 Tallinn, Estonia | gpsr.requests@easproject.com

Printed in China

10 9 8 7 6 5 4 3 2 1

Contents

Preface

As a geologist, I see everything through the lens of time. Nothing in nature is static; change is always present even if imperceptible in our life experience. If we view the natural world in the context of time, then all that we see becomes a story. I always tell people that geology is the art of storytelling. This is true at all spatial scales from a single rock or fossil specimen to rock outcrops, landscapes, and entire continents. Everything in nature has a backstory, and what we see is the way it is because of its history. Nothing can be truly understood without reference to time and change.

Getting a glimpse into the history of a landscape does not take away from its inherent beauty. For me, it only deepens the awe and amazement that I experience. It also causes me to want to look more deeply and closely at the natural environments around me. Catching clues to the geological and natural history of a landscape captures my attention whether I am hiking the Konza Prairie of Kansas, driving through Rocky Mountain National Park, walking within the volcanic caldera of Haleakalā in Maui, marveling at the beauty of the Rangitata Valley in New Zealand, hiking the West Highland Way in Scotland, or looking out my front door at home.

Landscapes are molded by a myriad of forces acting over a bewildering range of time scales. The motions of continents, the uplift and erosion of mountains, the rise and fall of sea level, and changing global climates create the large-scale context for the formation of a landscape. Superimposed on this are the actions of weathering and erosion processes as well as the impact of the native biological ecosystems that inhabit and transform a landscape.

Looking more deeply and closely also results in a growing recognition of the extraordinary diversity of nature. Taking the time to recognize the diversity of living things within a landscape transforms how we see that place. The prairie changes from just a sea of grass to a diverse ecosystem of grasses, flowering plants, insects, birds, mammals, and reptiles that interact in complex ways and change dynamically with the seasons. Similarly, recognizing the diversity, and environmental and ecological significance, of rocks brings to light a previously unseen record of the past.

Paying close attention to nature continually enlarges our appreciation and wonder. This photographic and geological tour of Kansas will touch on only a few aspects of its geological past, but I hope that it may stimulate your own exploration of your natural geological and biological environment as well as those you visit.

Introduction

The history of the Earth is dynamic. Though it may not be apparent during our short span of life, the surface of the Earth and the life it supports are in a constant state of change. These changes are both progressive (or directional) and cyclic. Earth history is also vast, with time scales that are hard to grasp. Geological time is often referred to as "deep time" to emphasize its distinction from time as seen from a human perspective.

Rocks provide our window into the deep time of Earth history. In Kansas, the rocks we see consist almost entirely of sedimentary rocks stacked in nearly horizontal layers. Sedimentary rocks are those formed by minerals chemically precipitated out of water or by sediment transported and deposited by wind, water, or ice. These sedimentary layers are like the pages of a book, and to read the story, you must know where to start. If we are reading rocks, we must read them from the bottom upward, from those sedimentary rock layers formed earlier to those formed later. This key to interpreting the rock record is referred to as the principle of "superposition": the rock layers or "strata," at the bottom of a sequence are older than those at the top. In addition, sedimentary rocks preserve evidence that enables us to reconstruct ancient environments. The mineral composition, texture, structure, and fossil content of these rocks reveal the environments in which they were originally formed. A vertical sequence of diverse sedimentary rock layers thus indicates a changing series of environments over geological time. It is a record of Earth history at that place.

Sedimentary rock layers are grouped into named units to aid in tracing, or "correlating," them from one locality to another. Stacks of sedimentary layers are subdivided into named "stratigraphic units" based on their shared mineral compositions, textures, and fossil content. These units can be easily recognized and mapped over a wide area. "Formations" are the primary units that are used to subdivide a stacked sequence of sedimentary rocks and can range from a few feet to hundreds of feet thick. Using these convenient handles makes communication easier and allows access to more detailed descriptions available in the geological literature. Geologists compare named stratigraphic units of the same relative age in different localities in order to reconstruct the geographic distribution of past environments.

The varied landscapes of Kansas are reflections of the underlying geology. Regions of similar topography are called "physiographic provinces" and are underlain by rocks of similar age and composition. In this book, we will tour these provinces, from those underlain by the oldest rocks to those exposing sediments deposited in the relatively recent past. This is a tour not just of the scenic beauty of Kansas but also of Earth history. Each region of Kansas is placed in the long chronology of Earth history, including changing sea levels and global climates, the uplift and erosion of mountains, and the evolution of life over hundreds of millions of years.

The Physiographic Provinces of Kansas

The Earth's surface can be divided into geographical regions that are characterized by a distinctive topography, specific exposed rock types, and particular geological structures. These regions are called "physiographic provinces." The geology of these provinces, together with current climatic conditions, produces unique visual landscapes and ecological environments.

The geology of each physiographic province is itself a consequence of global events, depositional environments, and changing climates of the distant geological past. Thus, each province preserves a distinctive history of the midcontinent as well as a distinctive landscape.

This photographic tour of Kansas is organized by physiographic province, from the regions underlain by the oldest rocks to those with the most recent rocks and sediments. In general, the bedrock becomes younger from east to west as the elevation of the landscape rises to the west. The rock layers to the west are stacked on top of the older layers exposed to the east. In keeping with geological storytelling, we will feature the landscapes of Kansas in chronological order, from those recording the oldest records of geological history preserved in Kansas to those recording more recent events.

The major provinces of Kansas can be defined by the ages of the underlying rocks. The simplified geological map of Kansas shows the distribution of rocks of different ages across the state. The oldest rocks exposed in Kansas occur in the southeasternmost corner of the state and were deposited during Mississippian time about 345 to 330 million years ago. The marine limestones occurring there represent the edge of the Ozark Plateau, which extends into Missouri, Oklahoma, and Arkansas. Occupying most of the eastern quarter of the state is the Osage Cuestas Province and the associated Cherokee Lowlands and Chautauqua Hills, composed of rocks deposited during the Pennsylva-

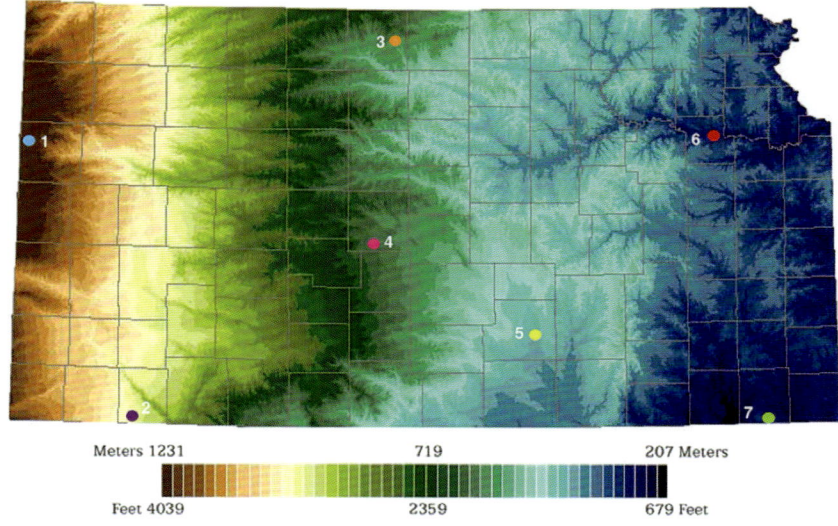

The elevation in Kansas rises gradually from east to west. The lowest point, at 679 feet above sea level, is where the Verdigris River exits the state into Oklahoma just south of Coffeyville in Montgomery County. The highest point, at 4,039 feet above sea level, is on the Colorado border in Wallace County. Although not a discernible rise, this high point is humorously known as Mount Sunflower. The map shows seven Kansas locations in relation to the changing elevations: (1) Mount Sunflower in Wallace County; (2) Liberal in Seward County; (3) the geographic center of the contiguous United States; (4) Great Bend in Barton County; (5) Wichita in Sedgwick County; (6) Topeka in Shawnee County; and (7) the lowest point in Kansas. (Used with permission of the Data Access and Support Center at the Kansas Geological Survey.)

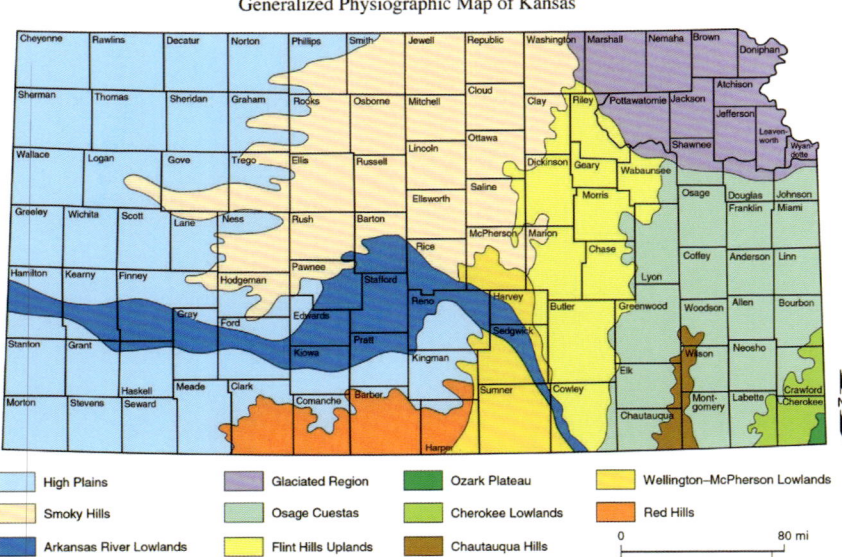

Generalized Physiographic Map of Kansas

High Plains	Glaciated Region	Ozark Plateau
Smoky Hills	Osage Cuestas	Cherokee Lowlands
Arkansas River Lowlands	Flint Hills Uplands	Chautauqua Hills
		Wellington–McPherson Lowlands
		Red Hills

0   80 mi
0   120 km

Based on the ages and types of rock exposed at the surface and their associated landscapes, this map shows the general areas assigned to the different physiographic provinces. These will be the context for our tour of the state. (Illustration courtesy of the Kansas Geological Survey.)

nian about 323 to 299 million years ago. These rocks display repeated cycles of marine limestone and black shale, alternating with fine-grained terrestrial rocks and river-deposited sandstone. Slightly tilted to the west, these rocks form the wooded ridges and flat plains characteristic of this province. Farther to the west, the Flint Hills Province is underlain by early Permian rocks deposited 299 to 280 million years ago. The repetitive alternation of nearly flat-lying resistant marine limestones and soft, fine-grained terrestrial rocks forms the terraced hills of the tallgrass prairie. To the south and west of the Flint Hills are the Wellington-McPherson Lowlands and Red Hills. This area is underlain by red, river-deposited sandstone and siltstone as well as gypsum and halite beds formed in the late Permian 280 to 260 million years ago.

After a long time gap of about 115 million years for which no surface rocks are present in Kansas, the Smoky Hills Province farther to the west preserves sandstones, shales, and chalks deposited during the Cretaceous 145 to 66 million years ago. The Monument Rocks that stand above the flat plains in Gove County and the badlands that dissect the plains are formed by the chalk beds of an ancient midcontinent shallow sea. The dry shortgrass prairie of westernmost Kansas belongs to the High Plains Province. Here, Miocene gravels, deposited from 23 million to 5 million years ago, cover the surface and rest on the eroded top of the Cretaceous chalks. The glacial and interglacial episodes of the Pleistocene Epoch, from 2.6 million to 11,700 years ago, left a record in the northeastern corner of Kansas, an area called the Glaciated Regions. This area was briefly covered by continental glaciers and is mantled by glacial sediments and fine windblown silt. Last of all, the Arkansas River Lowlands are covered by river channel and fertile floodplain deposits of the Arkansas River as well as vegetated hills of windblown silt.

The photos in this book will give you a taste of the surprising diversity of landscapes and environments in Kansas all rooted in their underlying geology. The past and present are inseparable, and each physiographic province has its own unique story to tell. The geological histories of these provinces have also shaped the history of the human societies that have lived on that land from millennia past to the present time.

This color-coded geological map shows the relative ages of rocks that are exposed at the surface in Kansas, from the oldest in the southeast corner of the state to the youngest in the west. Each color represents not only a particular time interval but also a unique set of rock types reflecting different environments at the time of their formation. Each physiographic province is thus defined by rocks recording a distinctive time in the geological history of Kansas. (Illustration courtesy of the Kansas Geological Survey. A more detailed geological map with updated nomenclature can be found at https://www.kgs.ku.edu/General /Geology/state23.html.)

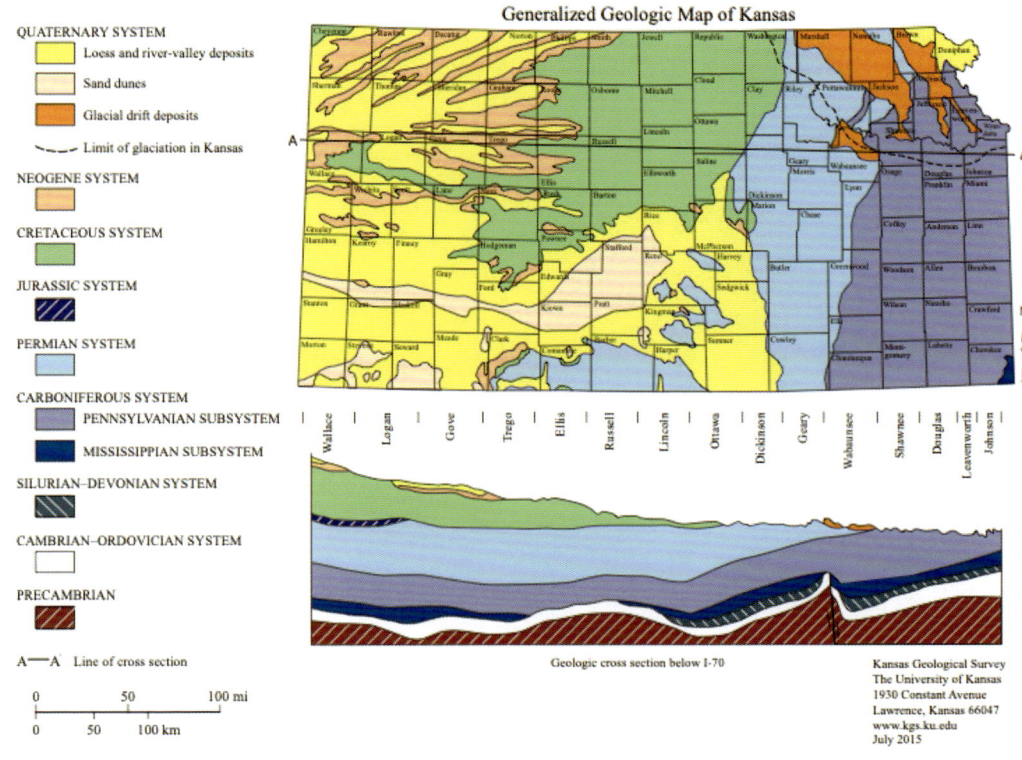

Generalized Geologic Map of Kansas

QUATERNARY SYSTEM
- Loess and river-valley deposits
- Sand dunes
- Glacial drift deposits
- – – – Limit of glaciation in Kansas

NEOGENE SYSTEM

CRETACEOUS SYSTEM

JURASSIC SYSTEM

PERMIAN SYSTEM

CARBONIFEROUS SYSTEM
- PENNSYLVANIAN SUBSYSTEM
- MISSISSIPPIAN SUBSYSTEM

SILURIAN–DEVONIAN SYSTEM

CAMBRIAN–ORDOVICIAN SYSTEM

PRECAMBRIAN

A——A′ Line of cross section

0   50   100 mi
0   50   100 km

Geologic cross section below I-70

Kansas Geological Survey
The University of Kansas
1930 Constant Avenue
Lawrence, Kansas 66047
www.kgs.ku.edu
July 2015

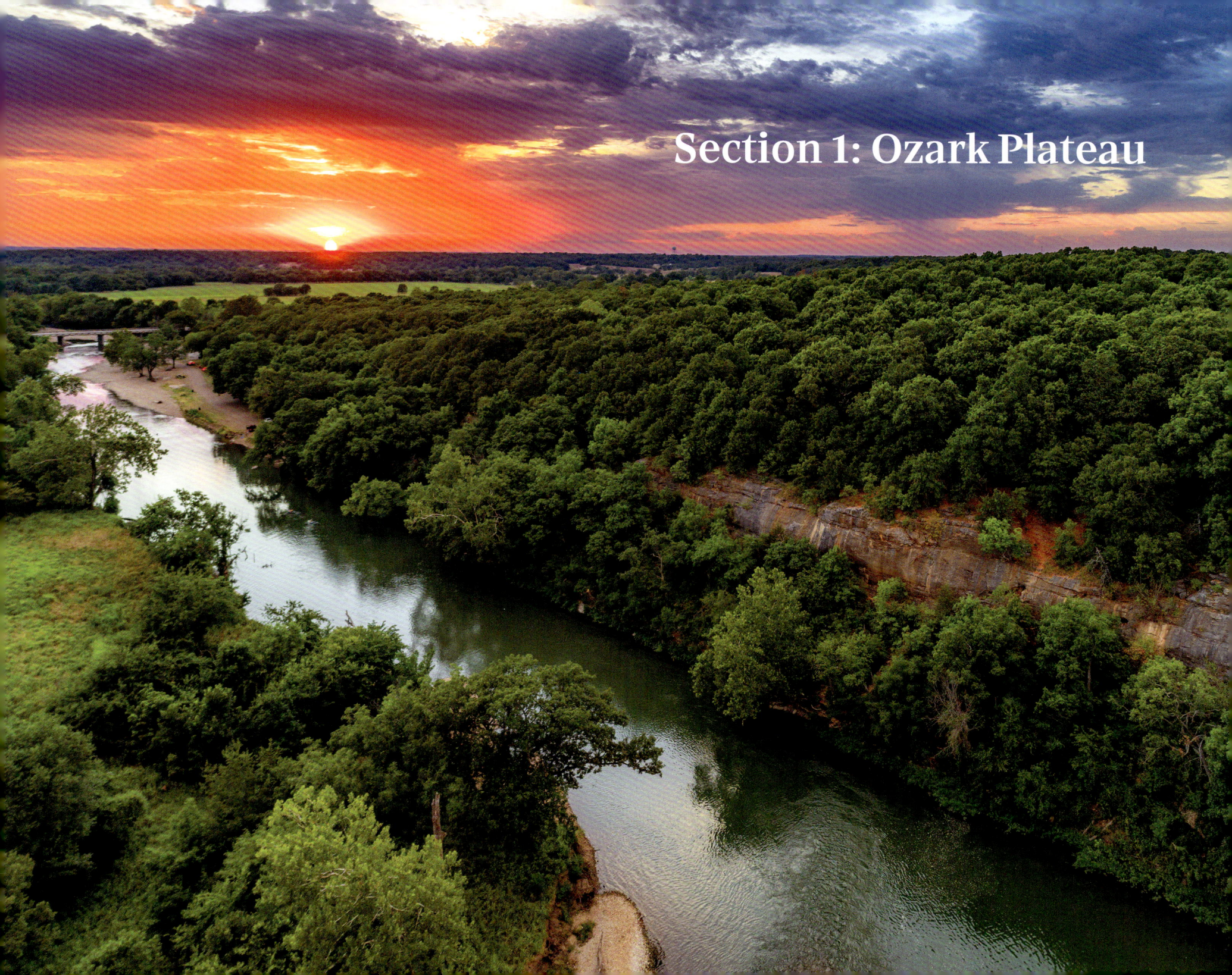

Section 1: Ozark Plateau

Rocks of the Ozark Plateau: Remnants of a Shallow Sea

Geology of the Ozark Plateau

The oldest rocks exposed in Kansas can be found in the southeasternmost corner of the state in Cherokee County. These rocks are primarily marine limestones containing abundant chert (or flint) as well as some silty green shales. Limestone is a sedimentary rock formed of calcium carbonate (calcite) that is primarily precipitated biologically by organisms as diverse as corals, clams, algae, and plankton. These sedimentary rocks were deposited during the middle to late Mississippian period around 345 to 330 million years ago. While occupying an area of only fifty square miles or so in Kansas, these rocks are just the western edge of a large uplifted area called the Ozark Plateau, which extends into Missouri, Oklahoma, and Arkansas and occupies nearly forty-seven thousand square miles. This western edge of the plateau is characterized by gently rolling hills that are interrupted by streams flowing within steep, rock-walled valleys. The area is the wettest in the state, with an average of over forty inches of rain a year. Forests of oak and hickory cover the hillsides along with flowering dogwood, sassafras, and hop hornbeam in the understory. This forest ecology is unique in the state. Some hilltops that are kept open by periodic fires are characterized by rocky desertlike areas called glades.

Previous page: Schermerhorn Park is a wonderful place to view the little corner of the Ozarks of Kansas. With a large cave, forested trails, and the beautiful Shoal Creek, this area has a lot to offer. I found a small cliff to the east of the park on Google Maps and sent my drone over to photograph it at sunset. The sun popped out for all of two or three minutes before disappearing behind the clouds again, creating this beautiful image. (Photo by Mickey Shannon.)

The combination of a wet climate and soluble limestone bedrock has resulted in widespread cave formation by the dissolving action of groundwater as it flows through fractures in the rock. As the caves are enlarged, the collapse of their roofs forms sinkholes as rock falls into the dissolved cavities below. As a result, most water flows below the surface within the dissolved cave systems, and surface streams are largely confined to the deep valleys bounded by limestone rock walls. This characteristic groundwater-formed landscape with caves, sinkholes, and sunken valleys is called karst. On the ground surface, the solution of limestone has resulted in the accumulation of hard, weather-resistant chert nodules as a layer of rubble.

The repeated exposure and weathering of the limestone over time resulted in sediment-filled fractures and cavities within the limestone. These cavities and fractures became the sites of the precipitation of lead- and zinc-bearing minerals by salty, sulfur-bearing fluids during the subsequent uplift of the Ozark Dome to the south and east. Galena is the primary ore mineral of lead. Its gray metallic appearance and cubic crystal shape make it easily identifiable. Galena and zinc-bearing sphalerite were mined in this corner of Kansas as part of the economically important Tri-State Mining District. Mining peaked in Cherokee County in the 1920s and declined after World War II, finally ending in 1970. The legacy of this mining includes collapsed mine shafts and surface subsidence as well as water contamination. Windblown dust from mine tailings and the use of tailings to pave roads resulted in high levels of lead contamination that have required the relocation of residents and intensive cleanup efforts.

The Record of an Ancient Ocean

The cherty limestones now exposed at the surface in the Ozark Plateau are buried by thousands of feet of sedimentary rock to the west and north. These late Mississippian limestones extend to Wyoming and Montana and beyond and are hidden now below much younger rocks. They record the existence of a vast but shallow sea that covered much of North America west of the Appalachians. Fossils present within these limestones include brachiopods, bryozoans, corals, and crinoids, all of which indicate shallow, well-lit, and oxygen-rich waters. The presence of land-derived sediments in the form of silty shales and evidence of erosion of the limestones also indicate times of exposure when the seas retreated. During the subsequent Pennsylvanian period, the collision of South America with North America created the Ouachita Mountains in Arkansas and southern Oklahoma and uplifted the Ozark Plateau to the north and east. This collision of continents began the assembly of the supercontinent called Pangaea.

Ozark Plateau Landscapes

Schermerhorn Park near the town of Galena provides one of the best localities to see the cherty limestones and karst landscape that characterize the Ozark Plateau in Kansas. The hard chert layers and irregular nodules stand out in relief on the weathered surfaces of the limestone.

The edges of a photo form a frame around the image that helps shape the way it is viewed. Elements in a photo can also form a frame within the scene. In this photo, the two trees on the sides frame the cherty limestone rocks at Schermerhorn Park in the background.

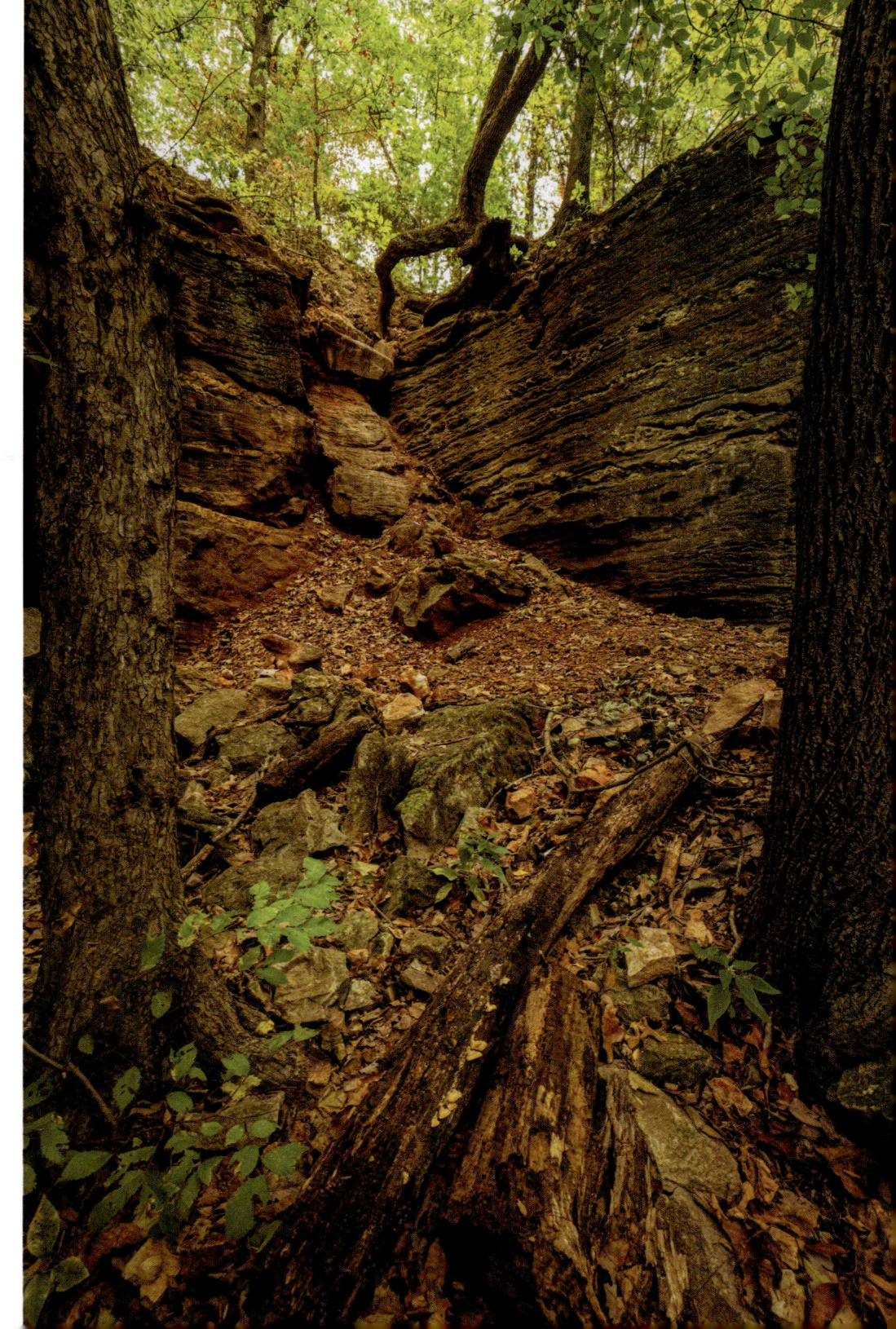

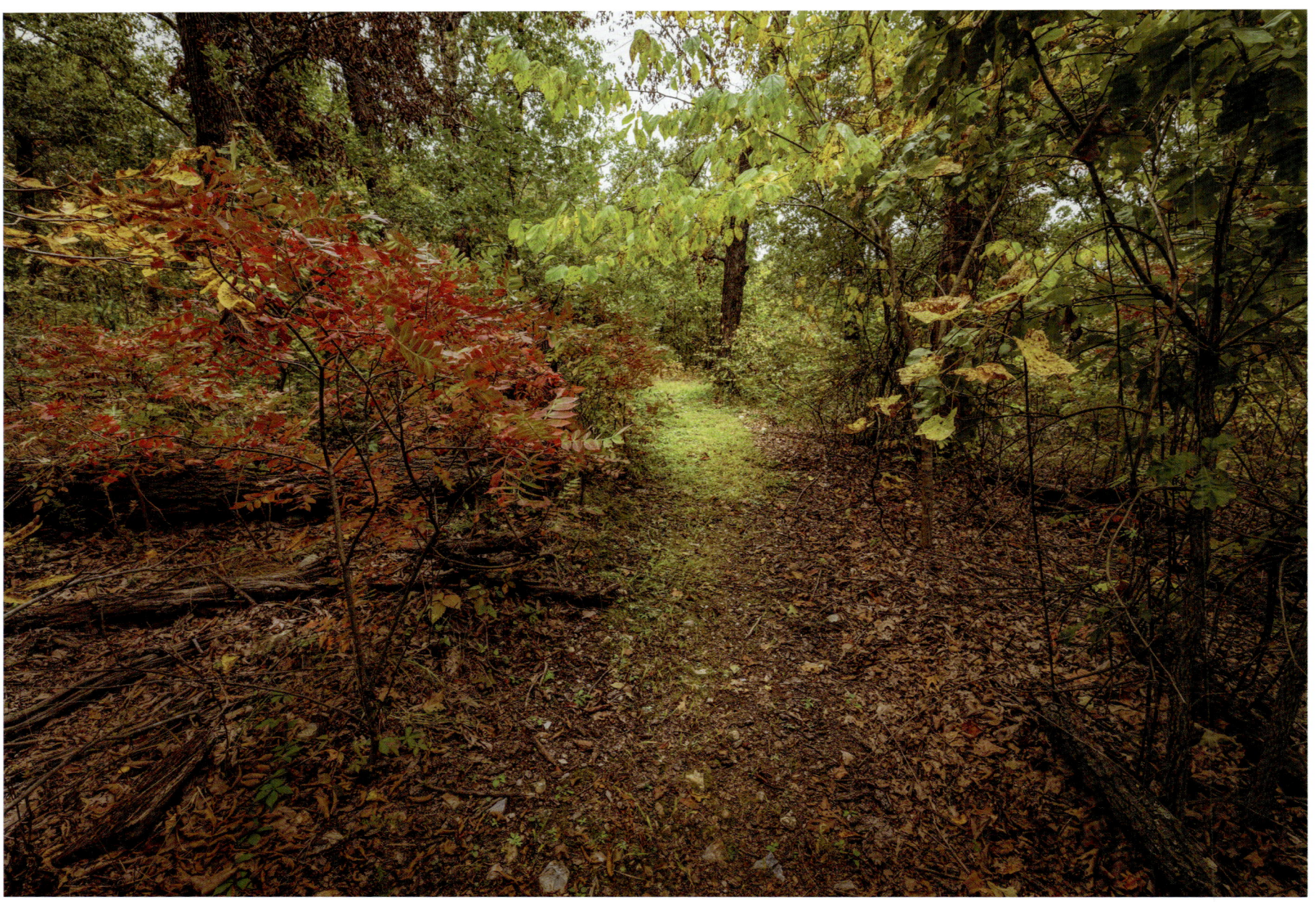

Opposite: Most of my photography is done in wide-open spaces. However, I enjoy walking on trails through wooded areas, especially during the fall. In addition to the quiet of trails like this one at Schermerhorn Park, wooded areas provide different colors and textures to photograph. The leaves on the trees just starting to develop their fall colors and the leaves on the ground contrast strongly with the green leaves and the green grass on the trail ahead. The downed tree limbs to either side seem to point the way forward.

Below: The deep black of the cave opening at Schermerhorn Park in Cherokee County immediately caught my attention. The limestone rock formed strong lines here, and I was happy that I could compose the photo to help draw the eye along the rocks to the cave, creating an even greater visual pull through the photo.

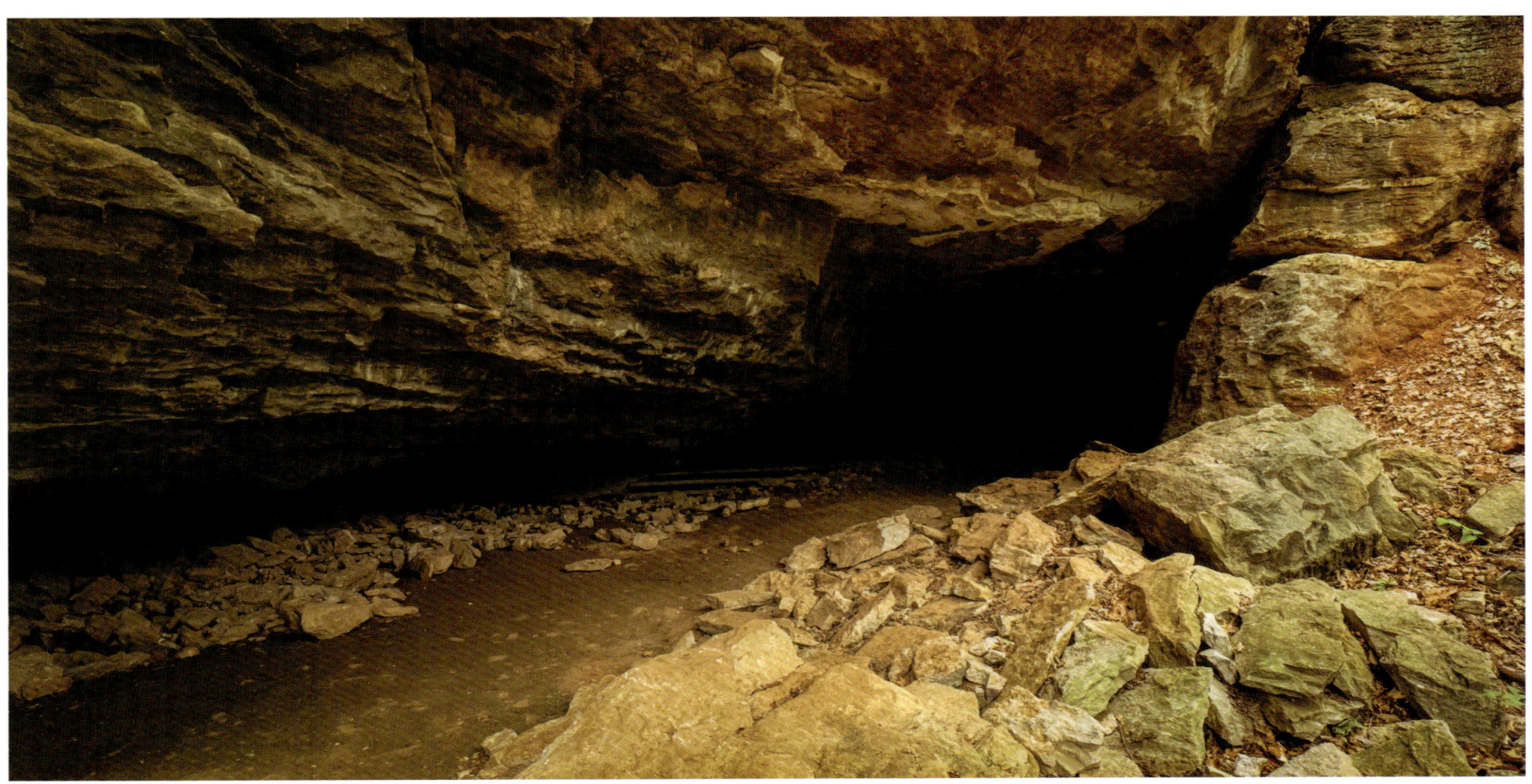

Fossils of the shallow-water invertebrates that lived on the seafloor can also be seen. In the park and along the nearby Shoal Creek, you can see the steep, rocky cliffs and valley walls that are typical of the Ozarks. Outcrops of weathered limestone are covered in dense forest dominated by oak and hickory. The area is also host to diverse bird species, attracting birders from across the state.

A special point of interest in the park is Schermerhorn Cave, with a large mouth from which flows an underground stream. This is typical of caves throughout the Ozark Plateau, where streams flow through cave systems below the surface. Schermerhorn Cave is home to bats, frogs, and a number of rare salamander species, and access to the half-mile-long cave is blocked to protect this important habitat.

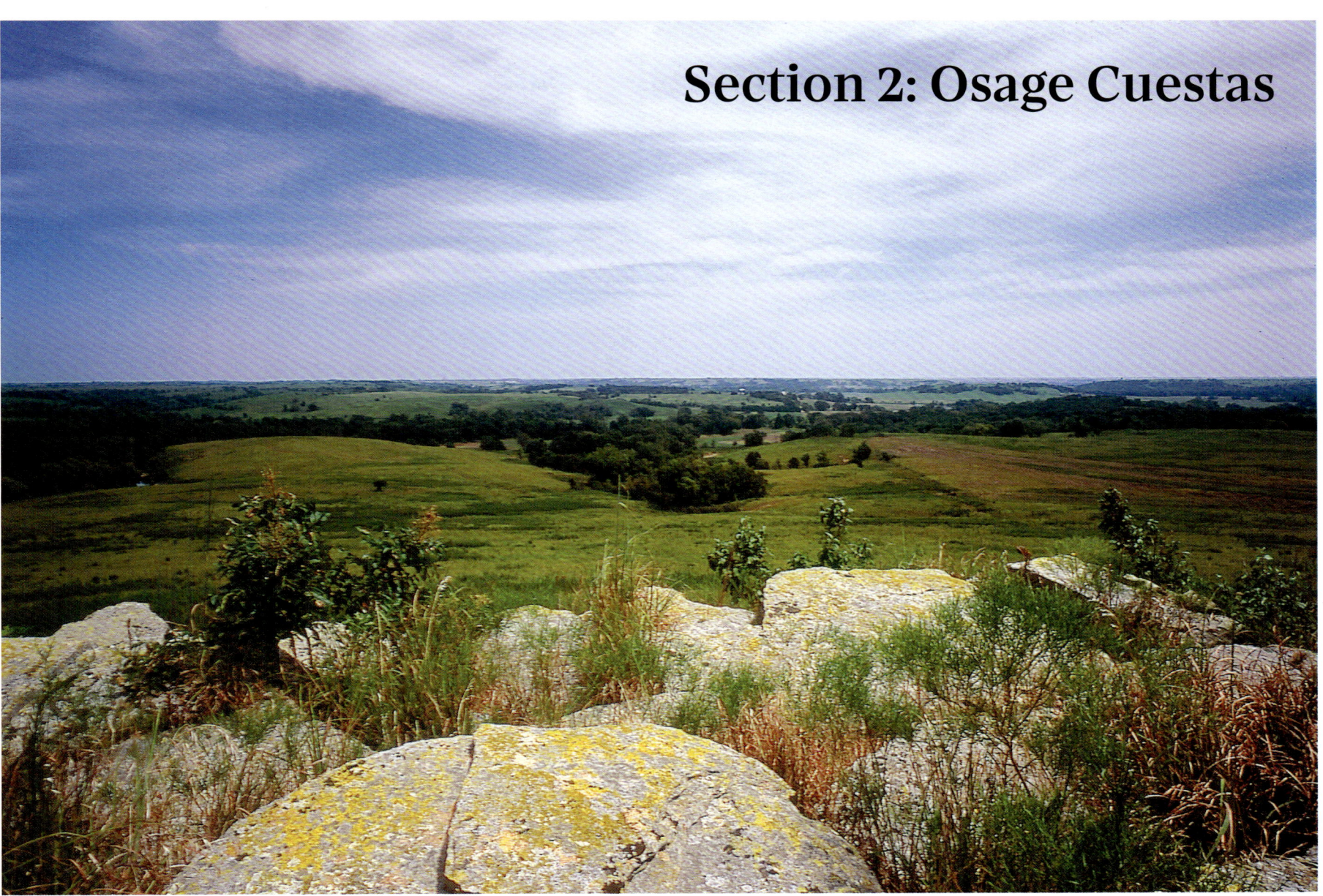

Section 2: Osage Cuestas

Rocks of the Osage Cuestas and Related Regions: Ancient Ice Ages and Peat Forests

Geology of the Osage Cuestas

"Cuesta" is the Spanish word for hill or cliff. The rocks in the Osage Cuestas region have been uplifted and tilted to the west forming ridges with steep eastern faces and gentle slopes to the west. The east-facing cliffs are capped by resistant limestone layers and can be up to two hundred feet high. The less resistant shales and sandstones between cuestas form relatively flat plains.

The limestones, sandstones, and shales exposed at the surface in eastern Kansas were deposited during the Pennsylvanian from about 323 million to 299 million years ago. These rocks occur in repeated cycles, with marine fossil–rich limestones and black shales overlain by terrestrial sandstones and shales deposited by rivers. Some of these rivers eroded deeply into the underlying marine units. Also formed during times when the area was exposed as land are ancient fossil soils, or "paleosols," and thin coal beds. These terrestrial rocks were subsequently drowned by rising sea level and overlain by marine limestones, beginning another cycle. These repetitive marine-to-terrestrial cycles are called "cyclothems" and are characteristic of Pennsylvanian-age rocks throughout the midcontinent in Illinois, Nebraska, Missouri, and Kansas.

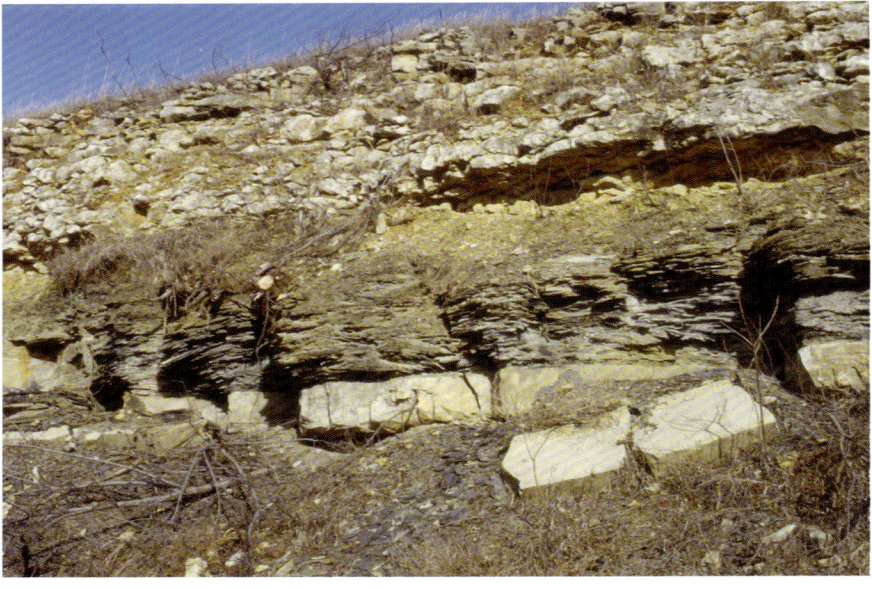

The marine phase of this late Pennsylvanian cycle begins with a thin limestone layer that represents the initial flooding of the land surface by the rising ocean. This is overlain by a finely layered black shale recording deeper water and low oxygen conditions at the seafloor. Above the black shale is a much thicker, fossil-rich limestone interval accumulated on a shallow seafloor. As sea level continued to fall, the limestone was in turn overlain by terrestrial shales, river-channel sandstones, and thin layers of coal. Another marine limestone at the top marks the next sea-level rise. (From bottom to top: Leavenworth Limestone, Heebner Shale, and Plattsmouth Limestone exposed three miles west of Lawrence along I-70. Photo by Keith Miller.)

Previous page: A view from the top of a cuesta in the Osage Cuestas Province looking to the east from Greenwood County. In the foreground can be seen the resistant limestone layers that form the tops of the east-facing cliffs. Beyond are gentle west-facing slopes formed by the backs of the tilted ridges. (Photo courtesy of the Kansas Geological Survey, John Carlton, photographer.)

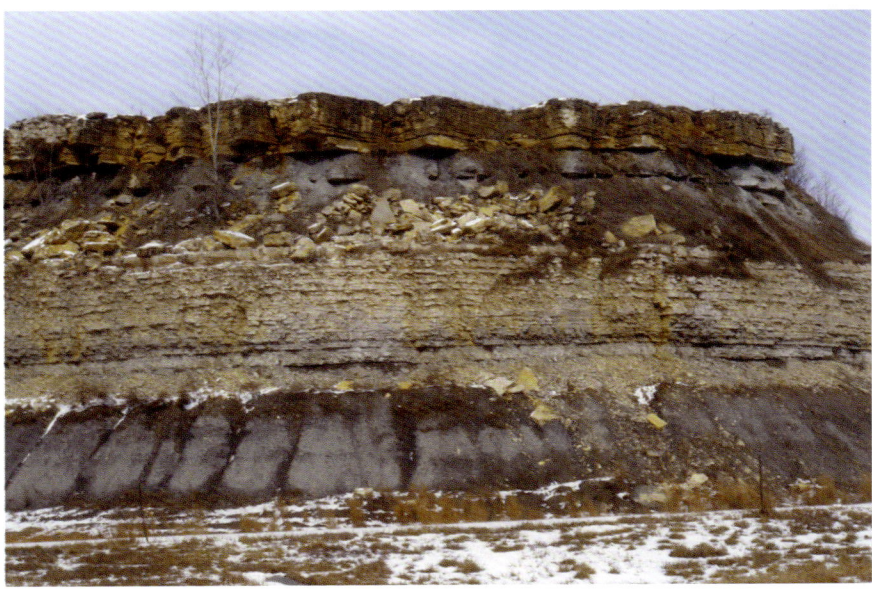

This exposure shows two terrestrial-to-marine cycles from the early Pennsylvanian. The terrestrial organic-rich gray shales record water-saturated conditions such as those present in marshlands and peat-forming environments. Thin coal beds representing ancient peat layers are also present within these mudrock intervals. The rubbly to thinly layered resistant limestones contain diverse assemblages of shallow marine fossils. (Lane Shale, Wyandotte Limestone, Bonner Springs Shale, and Plattsburg Limestone exposed at the Holliday Road interchange of I-435, Kansas City. Photo by Keith Miller.)

Related Regions of Eastern Kansas

The narrow band of rolling Chautauqua Hills runs through the center of the southern Osage Cuestas region. The thick sandstone and shale deposits of this area were deposited by large river systems that flowed into the ancient sea to the south. One of the largest of these, represented by the Tonganoxie Sandstone, had valleys up to twenty miles wide and ninety-five feet deep. These valleys were cut into the underlying rock units and filled with river-deposited gravel, sands, and associated muds. The valley-filling sediments were derived from the erosion of earlier sedimentary rocks to the east and north. These sediments were carried from as far away as the Appalachian Mountains, which were then being uplifted from the collision of Africa and North America as part of the assembly of the supercontinent Pangaea.

The Cherokee Lowlands is an area in the southeast corner of the state characterized by rolling hills to flat plains. This area represents the oldest Pennsylvanian-age rocks exposed in Kansas. In addition to river-deposited sandstones and shales, there are thick beds of coal. The coal layers, found mostly below the surface, were strip-mined from the late 1800s into the 1970s. Giant power shovels removed the overburden, creating trenches up to one hundred feet wide and one hundred feet deep. Many of the abandoned strip mines have since filled with water, and the land is now part of the Mined Land Wildlife Area in Cherokee and Crawford Counties.

Rocks as Windows into the Past

A closer look at the various marine and terrestrial rocks of eastern Kansas reveals a detailed record of changing environments and climates during the Pennsylvanian. The limestones are commonly fossil-rich with calcite skeleton-forming sponges, small branching corals, and bryozoans forming small columns and reef-like mounds. These limestones formed in warm, shallow water. The dark gray to black marine shales formed in somewhat deeper water that was low in oxygen and supported a very low-diversity community of organisms adapted to that environment.

The drainage pattern of the ancient Pennsylvanian river systems can be reconstructed from the channel sandstones. The channels extended southwestward from northwestern Missouri and then through Kansas toward the ancient ocean coastline to the south. The Tonganoxie Sandstone, mentioned above, preserves an interval of repetitive thin flood and ebb-tidal layers deposited at the upper end of an estuary opening toward the ocean. As the river approached the coast, rapid deposition occurred where the flow from the river interacted with the ocean tidal currents.

Other evidence indicates a gradual change toward a drier climate over the course of the Pennsylvanian. The large coal deposits of the Cherokee Lowlands were formed by extensive coal swamp forests in the earliest Pennsylvanian. These forests were dominated by long-extinct and somewhat alien-looking plants such as lycopsids, sphenopsids, and seed ferns. Terrestrial shales in eastern Kansas also preserve some of the earliest reptiles known in the fossil record. Lowland swamp forests occurred throughout the world at this time, and the resulting thick peat deposits have been preserved as coal beds. During the course of the Pennsylvanian, the coal deposits became much thinner and less extensive. The ancient soils (paleosols) also provide an important record of changing climate. The earlier paleosols indicate water-saturated soils in an ever-wet environment. Later in the Pennsylvanian, the paleosols show conditions more consistent with monsoonal or wet/dry seasonal climates. We will see that this drying trend continued through the overlying Permian Period recorded in the rocks of the Flint Hills to the west.

Causes of Pennsylvanian Cyclothems

You may be wondering what caused the dramatic and repeated alternations between the marine and terrestrial conditions described above. Why would the oceans rise and fall like a yo-yo? To answer this question, we must look

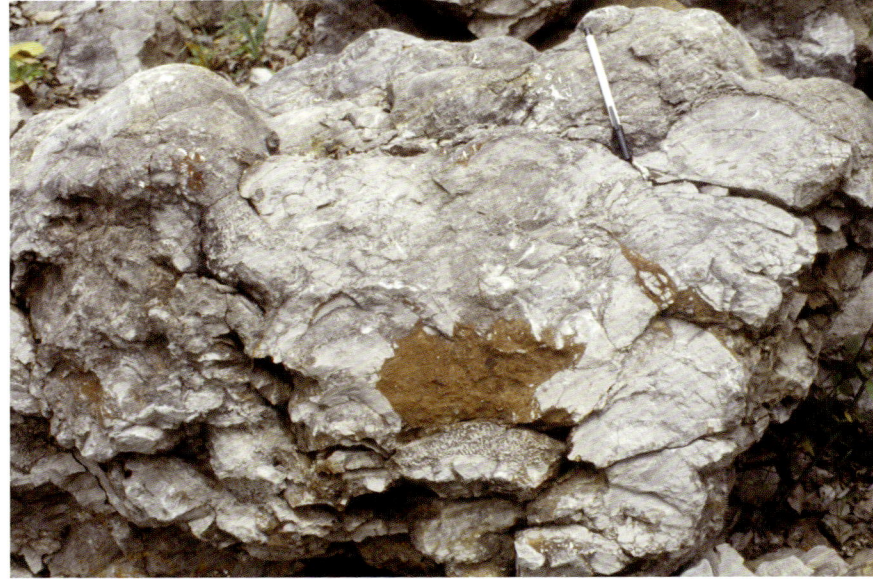

Some shallow marine limestones formed in a reef-like setting. This photo shows a fossil-rich limestone containing stacked hemispherical mounds, composed primarily of fossil sponges that constructed rigid calcite skeletons. These fossil sponges stand out against the reddish sediment surrounding them. Attached to the sponges are delicately branching corals, bryozoans, and crinoids. This marine limestone is over- and underlain by terrestrial shales and thin localized coal beds. (Upper Fort Scott Limestone [Higginsville Limestone] exposed in abandoned rock quarry near Girard, Crawford County. Photo by Keith Miller.)

elsewhere than the rocks in eastern Kansas, or even in the midcontinent of the United States. We must go to the continents of South America, Africa, Australia, India, and Antarctica, all of which were gathered around the south pole during this time. This southern part of the supercontinent Pangaea is called Gondwana. Here, we find evidence for a great period of continental glaciation—an ice age—during the time represented by the Kansas cyclothems. Such glacial periods are characterized by the repeated advance and retreat of the continental ice sheets. Each time the great glaciers advanced, the sea level would fall. Water that evaporated from the sea would fall as snow and be trapped in the glacial ice rather than returning to the sea. Conversely, each time the glaciers retreated and melted, water flowed back into the sea, causing the sea level to rise. Thus, the cycles observed in the rocks of Kansas record global sea-level changes tied to glacial and interglacial periods on continents half a world away.

OSAGE CUESTAS LANDSCAPES

Waterfalls can reveal geology that is otherwise covered and hidden under soil and vegetation. The waterfalls in eastern Kansas are typically capped by marine limestone units that are more resistant than the softer terrestrial shales above and below. Bourbon Falls exposes a typical Pennsylvanian cyclothem with the marine Hertha Limestone and underlying terrestrial Tacket Shale. At Sedan Lake, the thick, rubbly Plattsmouth Limestone contains abundant marine fossils and records shallow, well-oxygenated ocean conditions that followed the somewhat deeper, oxygen-poor seafloor conditions represented by the black shale of the Heebner Shale below. The Cedar Lake Falls displays the Stanton Limestone, which is famous because fossils of *Petrolacosaurus*, the earliest known diapsid reptile, are found only in a mudstone that filled a channel eroded into the Stanton Limestone in Garnett, Kansas. Such primitive diapsid reptiles are ancestral to all lizards, snakes, crocodilians, dinosaurs, and birds.

Spring flow was in full effect at Bourbon Falls in the spillway of Bourbon County State Lake. Flowers were beginning to sprout up near the falls as well. (Photo by Mickey Shannon.)

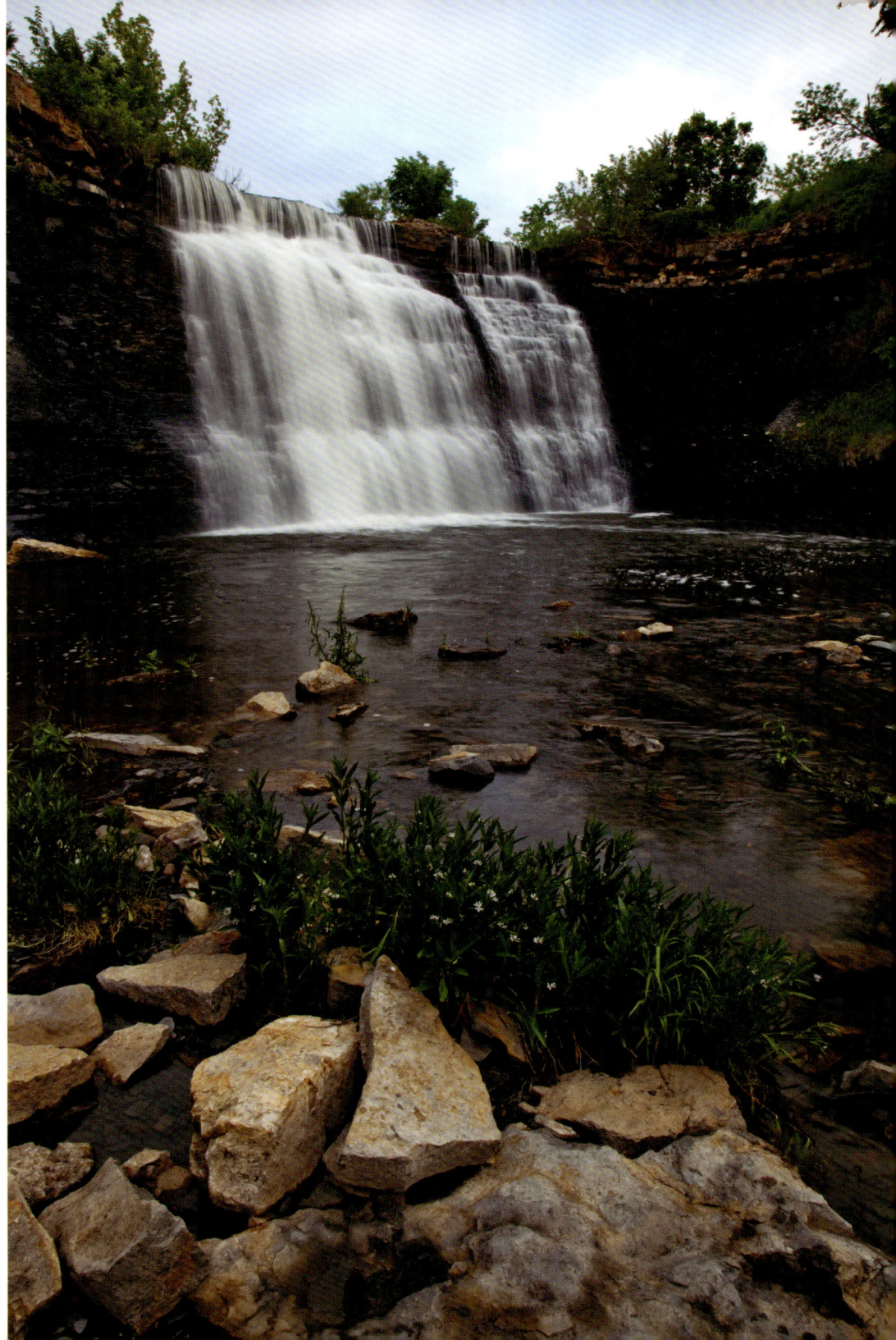

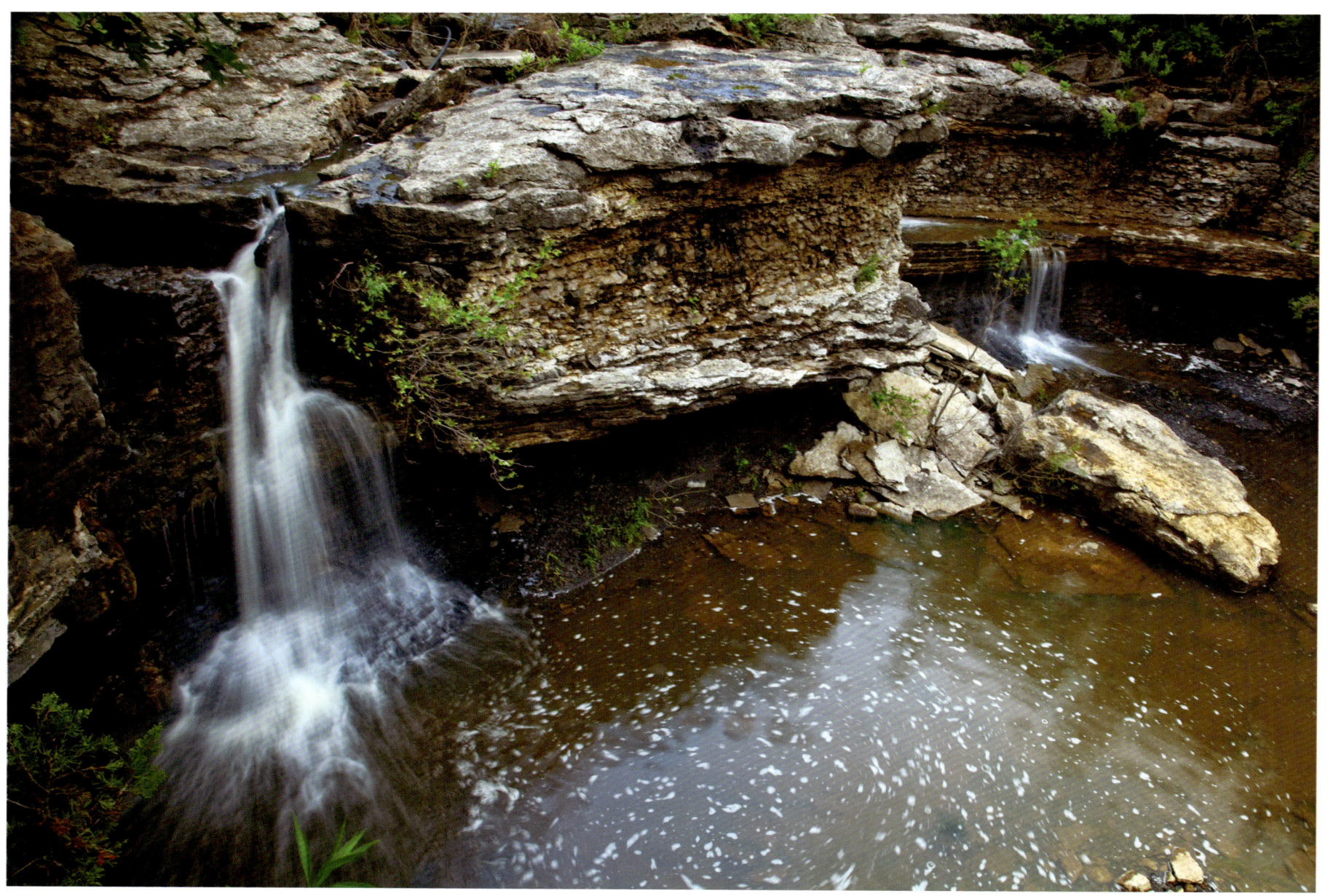

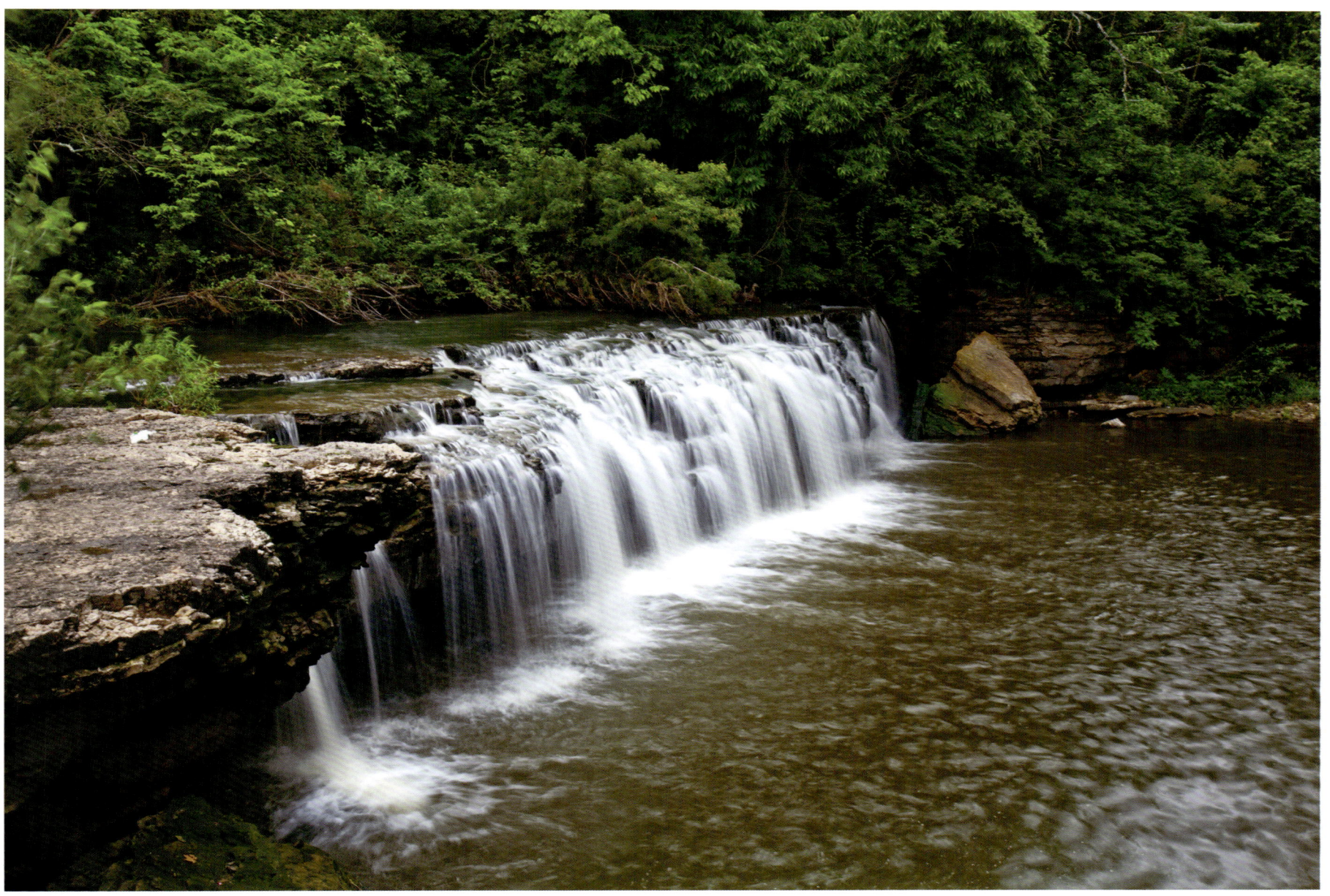

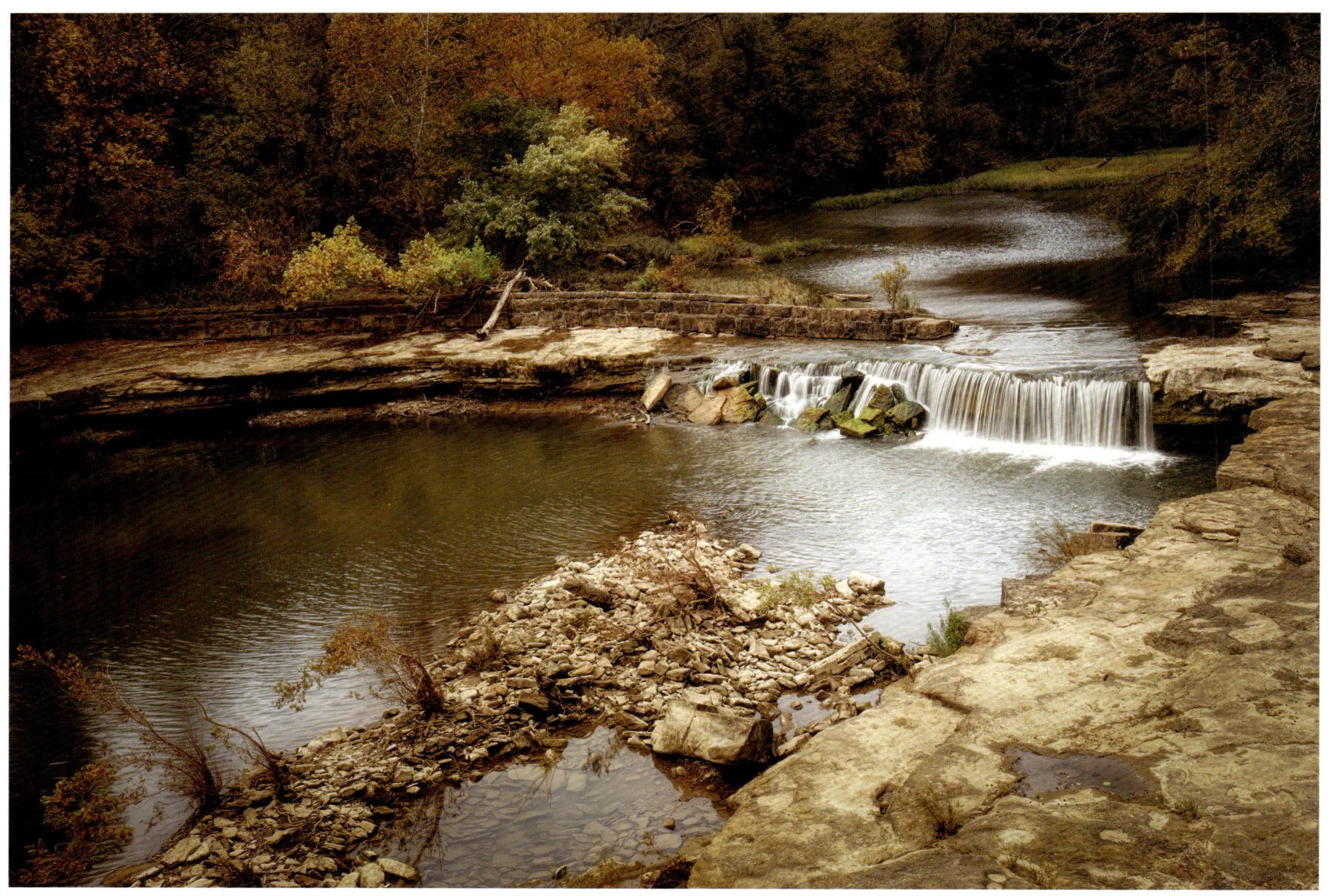

The sandstone cliffs above Mission Creek at Echo Cliff Park in Wabaunsee County provide great views of river-deposited sandstones that filled a series of channels that had been cut deeply into the underlying marine limestones during a time of sea-level fall, when the land was above sea level. This sandstone is informally known as the Indian Cave Sandstone and is up to seventy-five feet thick. The close-up image shows the fine horizontal and inclined laminations typical of deposition in river channels. The inclined laminations, or cross-beds, are produced by migrating ripples and record the strength and direction of the river current at the time they were formed. Such "sedimentary structures" preserve much important information about ancient environments.

Page 16: The upper falls at Sedan Lake near Sedan in Chautauqua County, Kansas. During heavy rains, these falls change from this gentle flow to a surge of water. (Photo by Mickey Shannon.)

Page 17: Earlier in my photography career, I was on a mission to photograph as many waterfalls in Kansas as I could. On this particular trip, I concentrated on the northeast section of Kansas, visiting a number of spillway waterfalls. One of my favorites was this one at Cedar Lake in the Kansas City region. With overcast skies, the deep colors of the trees contrasted with the water in motion below. (Photo by Mickey Shannon.)

Opposite: Elk River Falls in Elk County is seen here on a beautiful October day. The bridge near this waterfall provided a unique vantage point and the chance to take a photograph that captured the entire setting of the area and not just the waterfall itself. The fall weather, changing colors, and sounds of the waterfall combined to create a very peaceful moment.

Right: This close-up of Echo Cliff shows the texture and internal structures of this sandstone that are characteristic of river deposition. (Photo by Mark Clarke.)

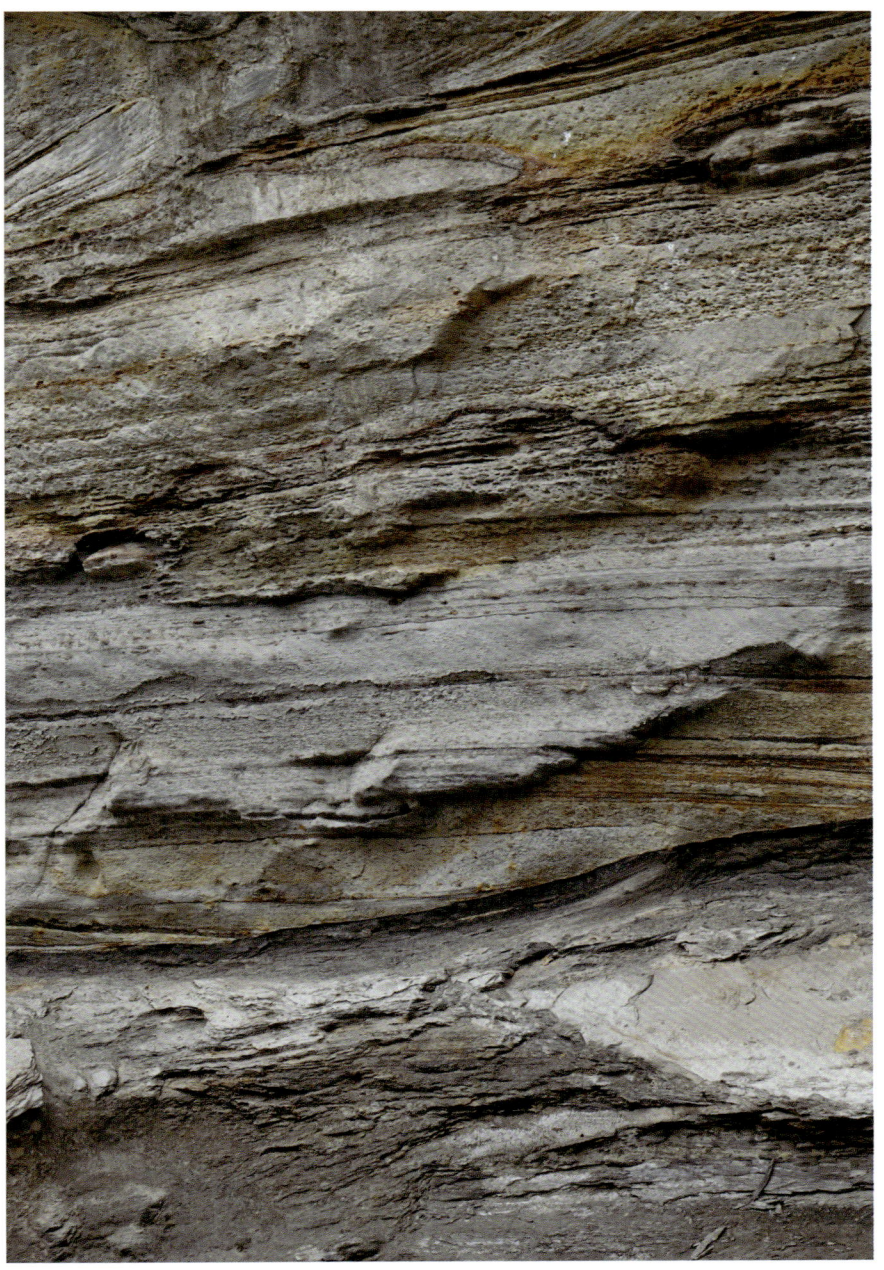

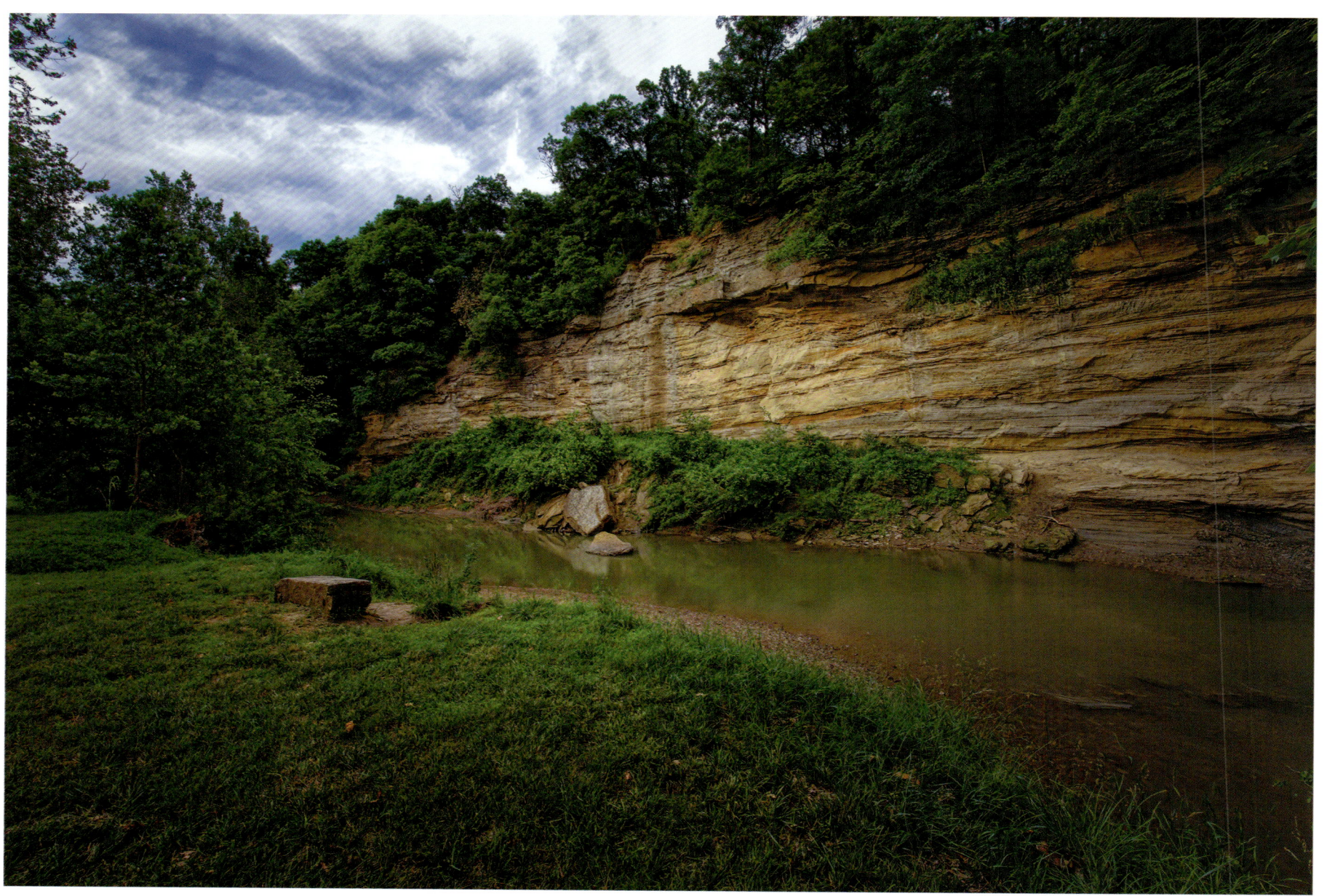

Opposite: The contrast between the colors of the trees and grass and the exposed rock initially caught my eye in this photo from Echo Cliff, but then the lines and the textures of the rocks held my attention.

Below: I photographed this image of Toronto Lake at Cross Timbers State Park in Woodson County a number of years ago, when I was still building a portfolio of images from Kansas state parks. The sky wasn't superinteresting, but the trees had started to turn to fall colors. While there was still some green, I wanted to send up the drone and see what it looked like from above. I loved the shimmer of the setting sun over the lake and took a four- or five-shot panorama to showcase the entire horizon to the west from above the lake. (Photo by Mickey Shannon.)

River-deposited sandstones and tidal sediments fill large river valleys that were cut deeply into the rock layers below during times of low global sea level. These large river systems flowed into the sea to the south. The thick sandstone units produced by these large rivers characterize the Chautauqua Hills. The rolling hills are capped by these resistant sandstones, which have created a unique terrestrial ecology called the Cross Timbers ecosystem, consisting of hardwood forests intermixed with prairie grasses. The forests of blackjack oaks, post oaks, and other hardwoods are well adapted to the sandstone bedrock environment. This ecosystem extends southward through Oklahoma and into north-central Texas.

The strip mining of large volumes of coal from the Cherokee Lowlands in Cherokee and Crawford Counties resulted in deep trenches that cut across the landscape with piles of strip-mined waste between. The deepest trenches

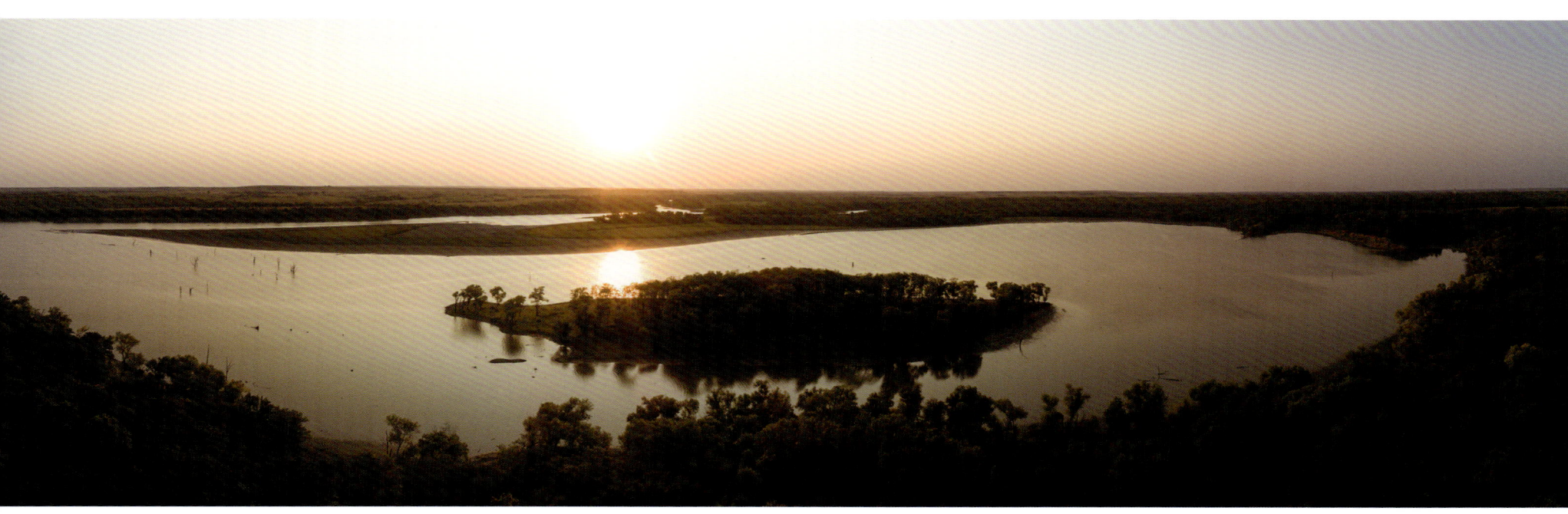

were dug in Cherokee County in the southeast corner of the state. One of the world's largest power shovels at the time, named Big Brutus, worked the strip mines there from 1963 until 1974. After the mines closed, the trenches filled with water and plants overgrew the piles of mine waste. Although companies were required to reclaim mined land after 1969, the older strip-mine lakes became part of the Mined Land Wildlife Area, which covers 14,500 acres, of which about 1,500 acres are public waters. The lakes are stocked with fish, and the surrounding mine waste is now covered in native grasses, woody shrubs, hickory forests, and wetlands, hosting a diversity of mammals and birds.

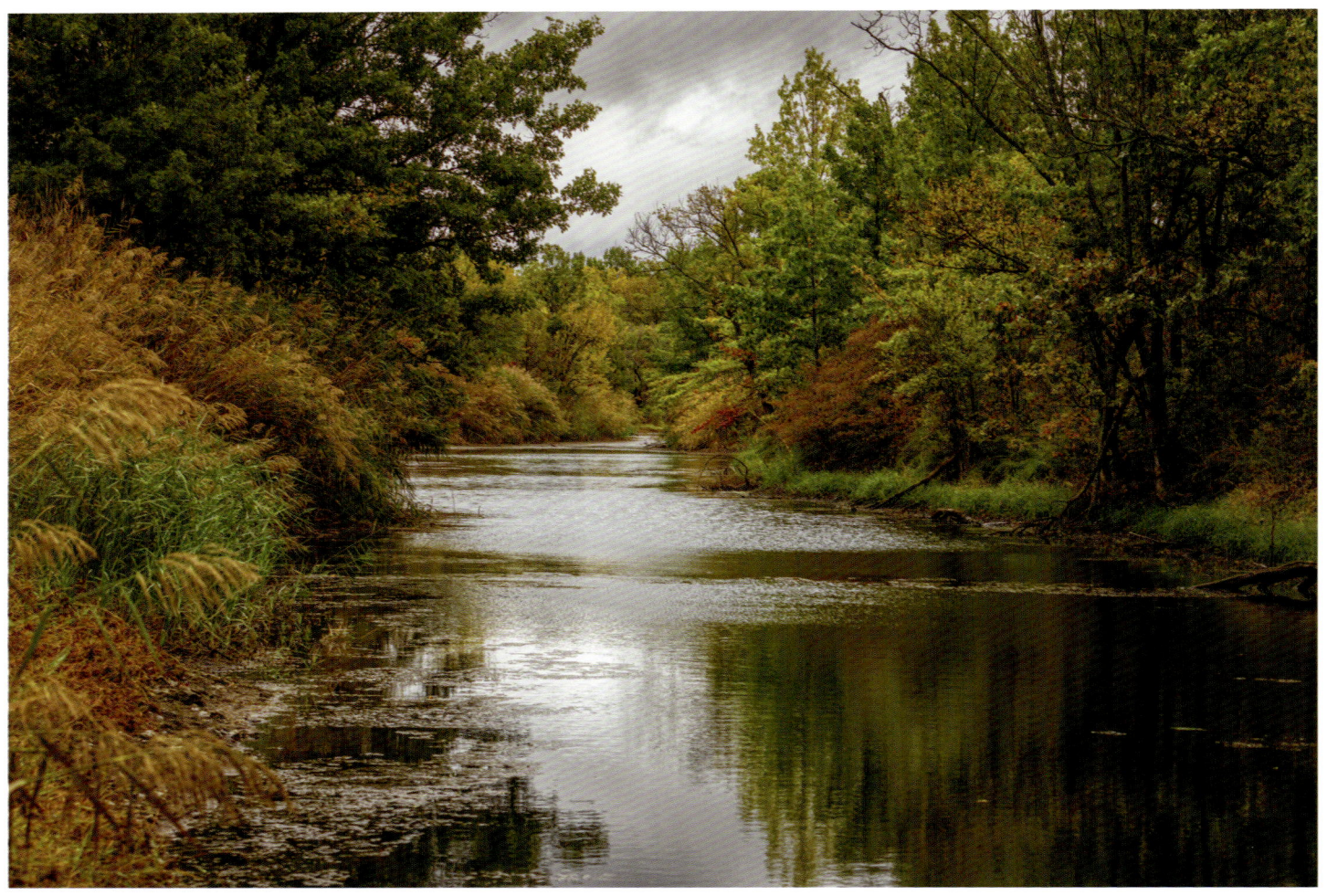

Left: The different layers of texture and color in this view down one of the waterways in the Mined Land Wildlife Area drew my eye. The fall colors were just starting to appear when I visited this area, giving me a wide range of subjects to photograph.

Opposite: Overcast conditions will often help bring out colors in a landscape. The clouds were just starting to break, letting in some light, when I made this photo. But the day was still overcast enough to really capture the brilliance of this view from the Mined Lands Wildlife Area.

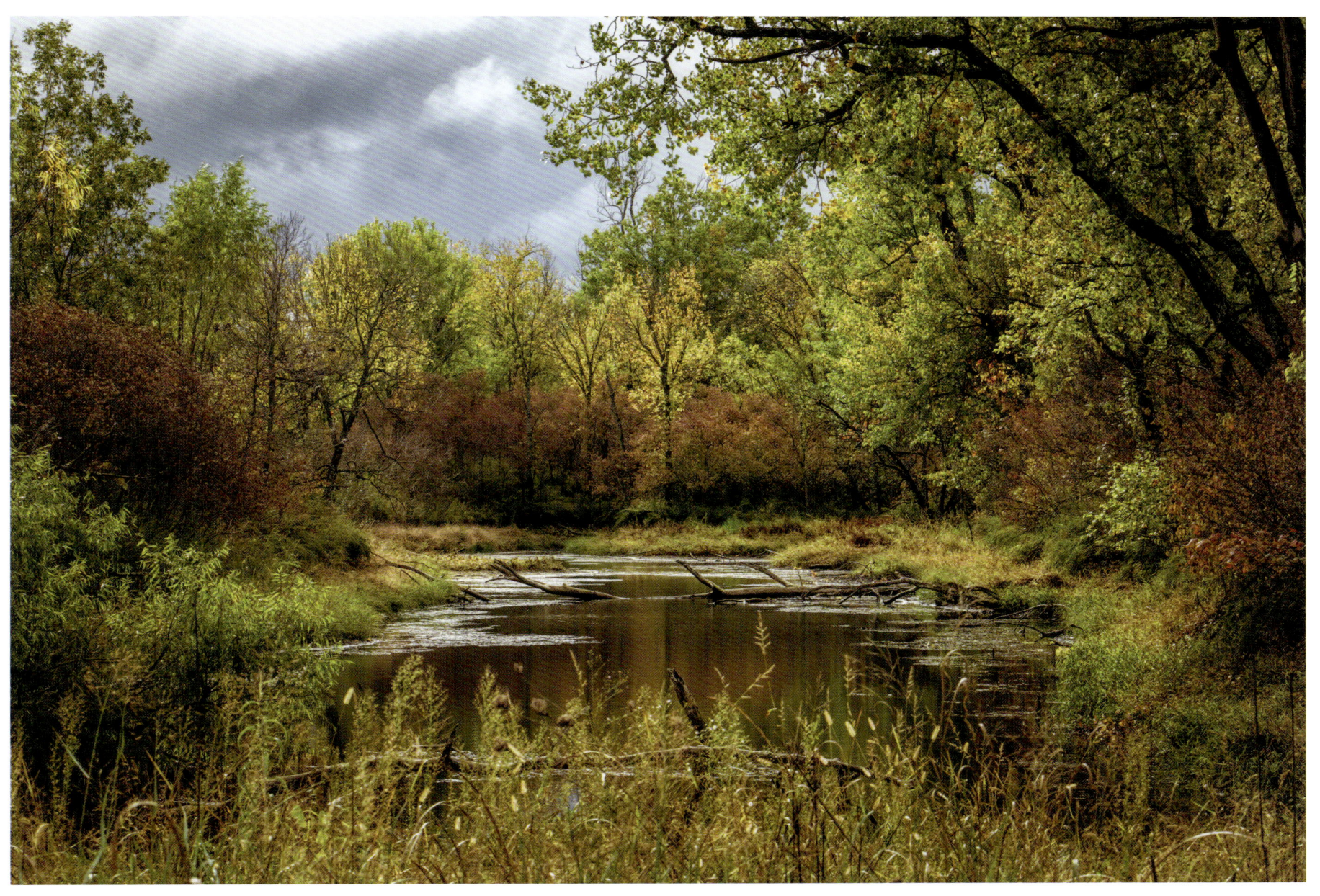

Section 3: Flint Hills

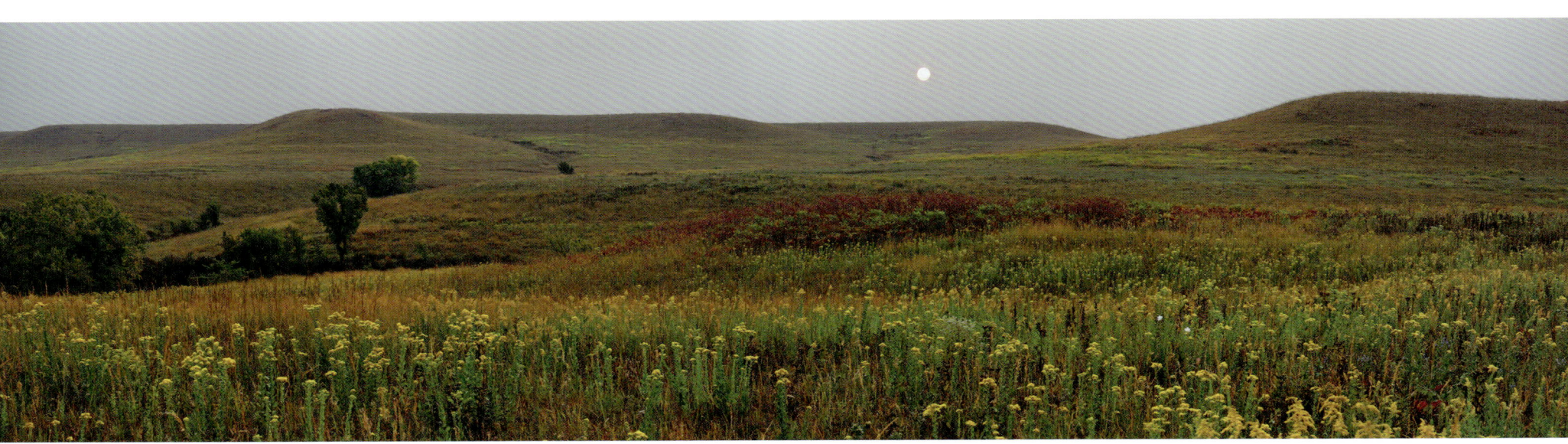

Geology of the Flint Hills

The sedimentary rock layers of the Flint Hills of Kansas record a period of time extending from the latest Pennsylvanian Period through the end of the early Permian. This encompasses a time interval of roughly 20 million years, from somewhat before 300 million years ago to about 280 million years ago.

The sedimentary rocks of the Flint Hills can be broadly classified into two types. The resistant layers are carbonate rocks (limestones and dolomites) that occur interbedded with thin, finely layered dark gray shales. These carbonates and gray shales are typically quite fossil-rich. Some limestones contain abundant wheat-grain-shaped fusulinid forams (a type of calcite-shelled protist) as well as diverse brachiopods, bivalves, crinoid segments, and sea urchin spines. The much less resistant reddish and greenish intervals are mudstones and siltstones. All of these rock units are laid out in nearly horizontal layers in the Flint Hills. The repeated alternation between the carbonates and colorful soft mudstones has produced the characteristic terraced landscape of the Flint Hills area, with the carbonates forming the tops of terraces and the mudstones forming the slopes. The most prominent characteristic of the Flint Hills geology is this cyclic repetition of rock types. These cycles are a continuation of the cyclothem pattern that characterizes the Pennsylvanian-age Osage Cuestas region to the east. The early Permian cycles have been widely traced throughout the Flint Hills region from Nebraska to Oklahoma.

Previous page: The Flint Hills come alive with color in the fall with the rusty brown of the tall grass, the bright red leaves of the smooth sumac, and the brilliant yellows of the goldenrod. When this photo was taken in Pottawatomie County, smoke from wildfires in the western United States was partially obscuring the sky, turning the sun into an eerily subdued point of light. (Photo by Mark Clarke.)

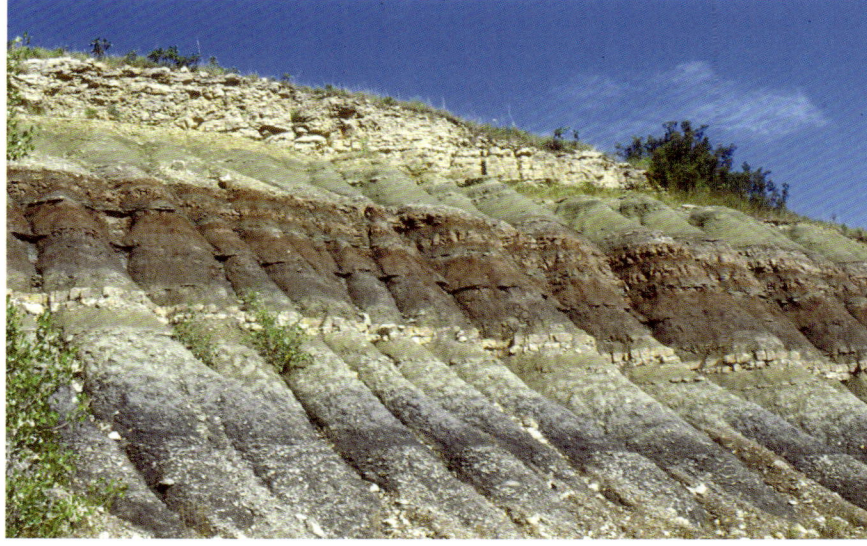

This roadcut exposure beautifully displays the characteristic sequence of rock types in an early Permian cyclothem. In this record of oscillating sea level, the thin, buff-colored hard layer in the middle and the thick, rubbly-looking layer at the top are limestones deposited in shallow ocean water. By contrast, the colorful banded mudstones and shales between record terrestrial conditions in a coastal plain setting. The distinctive colors and textures of these softer rocks reflect soil formation in varying climatic conditions. (Blue Springs Shale and Florence Limestone from the Seth Childs Boulevard [K-113] roadcut in Manhattan, Riley County. Photo by Keith Miller.)

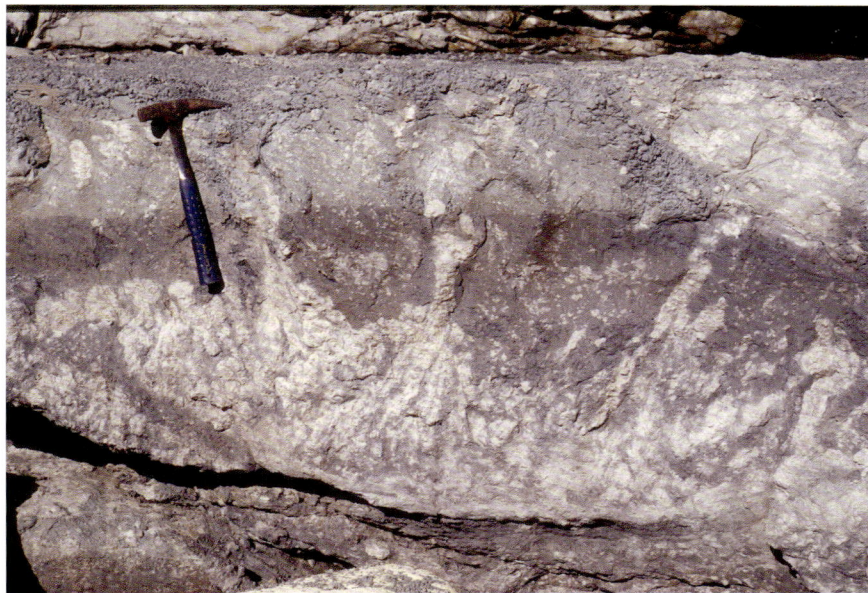

This mottled red mudstone represents a fossil soil, or paleosol. The light-colored features are carbonate concretions that formed around the roots of vegetation growing on the ancient land surface. Such accumulation of calcium carbonate nodules in soils occurs in semiarid or subhumid climates. The mottled reddish colors indicate that the soil was dry for much of the year. (Blue Rapids Shale from the Tuttle Creek Reservoir spillway near Manhattan, Riley County. Photo by Keith Miller.)

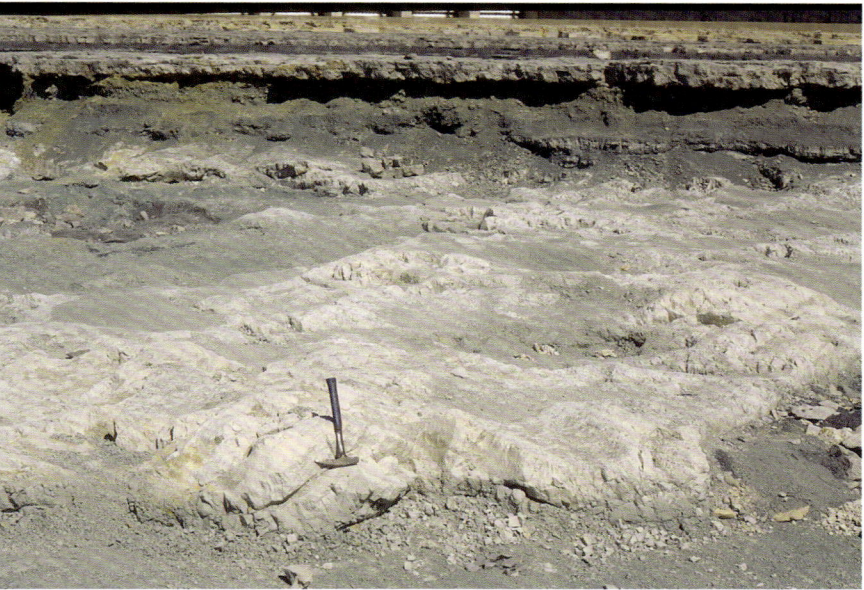

Resting above the red, nodular paleosol, this greenish-gray mudstone represents a distinctive type of paleosol characterized by its color and by undulating surfaces within the soil. Such features form by the expansion and contraction of clay-rich soils in wet/dry monsoonal climates. These paleosol types occur at the top of terrestrial intervals where they were abruptly overlain by limestones formed after the land was flooded by rising sea levels. (Top of the Blue Rapids Shale from the Tuttle Creek Reservoir spillway. Photo by Keith Miller.)

Many of the carbonate units contain abundant flint (or chert) nodules, which in some cases merge to form thin layers. Flint nodules are composed of microcrystalline silica that forms chemically by the replacement of the limestone in which the nodules occur. They literally grow inside the fossil-rich limestone, and remnants of fossils can often be seen within the nodules. It is these hard, weather-resistant nodules that give their name to the area. As the enclosing carbonates are slowly removed by weathering, the flint nodules are left behind as a layer of rubble on the surface. Flint is hard and will hold a sharp edge; it was used for thousands of years by native peoples for a wide variety of tools and weapons. These hard nodules also later prevented the early settlers from tilling the uplands—thus also preserving the native tallgrass prairie for later generations.

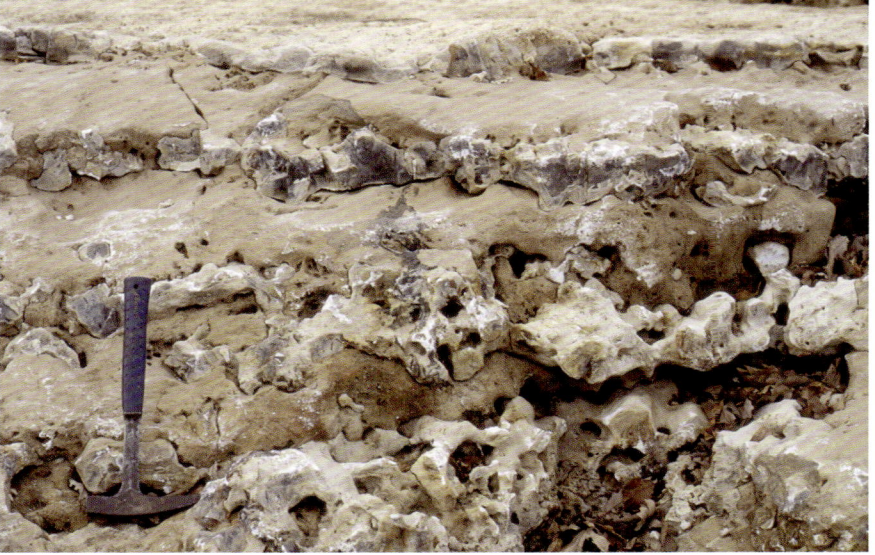

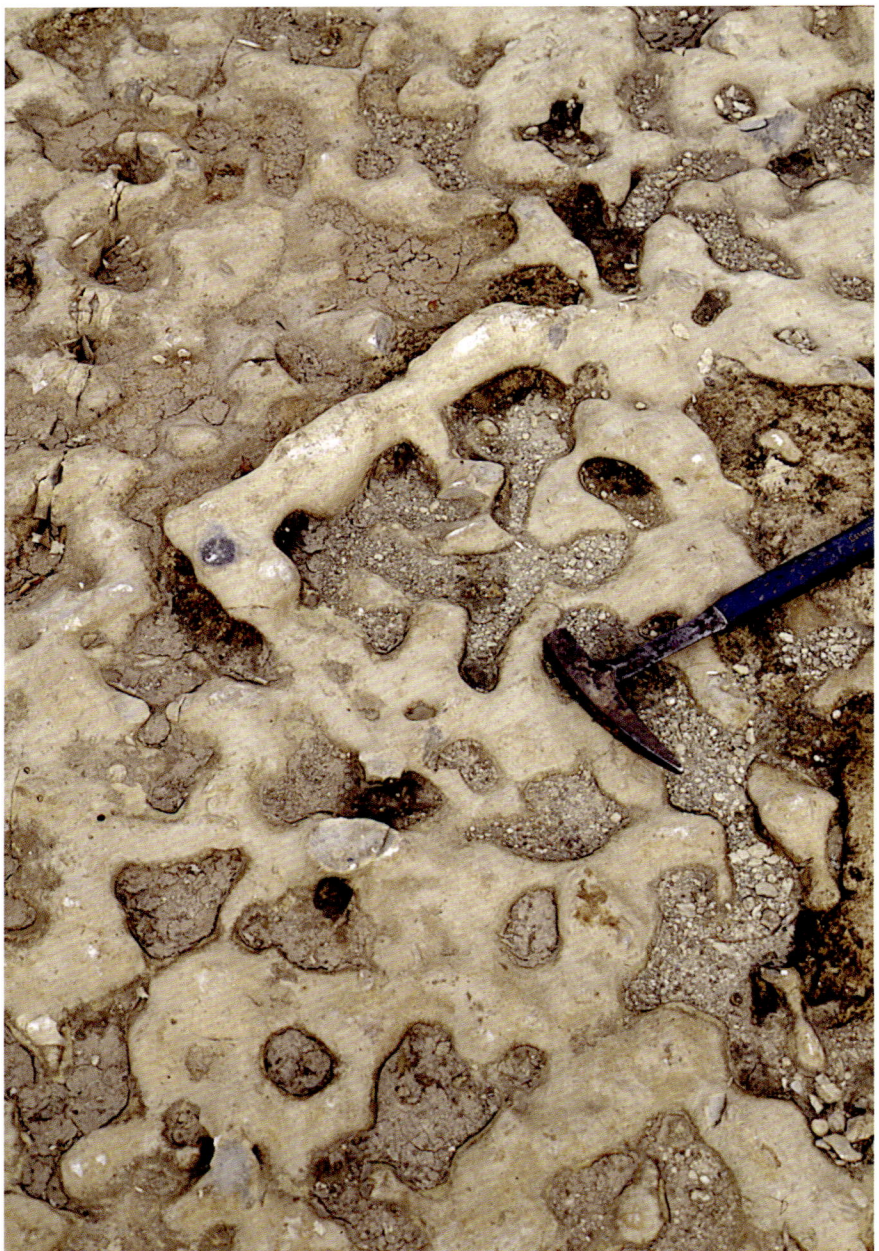

Opposite and above: These two close-up photos show flint nodule layers within the Florence Limestone. The irregular shapes of some of these nodules likely formed by the silica replacement of invertebrate burrow systems on the seafloor. Individual fossils also have been replaced by silica, resulting in beautifully preserved three-dimensional fossil specimens. The resistant flint nodules erode out of the limestones, forming layers of flint rubble on the ground surface. (Florence Limestone exposed in the eroded spillway of Milford Lake near Junction City, Geary County. Photos by Keith Miller.)

What Can the Rocks of the Early Permian Cyclothems Tell Us?

The early Permian cyclothems in Kansas typically begin with limestones and gray, fossil-rich shales that were deposited in a shallow sea. These limestones commonly culminate in layers recording intertidal environments and eventual exposure to the air. Variously colored mudstones lie above these shallowing marine rocks and show evidence of repeated extensive soil development. This interval of vertically stacked ancient fossil soils, or paleosols, is then sharply overlain by another marine limestone, and the cycle begins again. In the early Permian, these cycles are commonly twenty to thirty feet thick. Dozens of such alternations occur within the late Pennsylvanian through early Permian time period. The midcontinent was thus subjected to repeated flooding by shallow seas, followed by exposure as dry land.

The individual cyclothems of the Flint Hills also show consistent internal patterns that record repeated cycles of climate change occurring in step with changes in sea level. Many of the carbonate beds that formed in very shallow marine and coastal environments show evidence of arid climate conditions, such as replaced gypsum nodules (often present as quartz or calcite-filled geodes), salt-crystal molds, and finely layered dolomites. Overlying these arid coastal sediments are reddish fossil soils that typically contain calcite "caliche" nodules. The calcium carbonate that forms caliche nodules accumulates in modern soils where the climate is relatively dry. Above the reddish paleosols are found the very different-looking greenish paleosols that record yet another ancient environment. Soils similar to these form today in highly seasonal wet/dry, or monsoonal, climates.

In summary, what we see within each of the many stacked cyclothems is first a change from a shallow sea to intertidal conditions under rather arid climates. This is followed by the continued lowering of sea level and the exposure of the area as land—first under a relatively dry semiarid climate and then under a seasonally wet/dry monsoonal climate. The story of each of these cycles ends with the dramatic flooding of the vegetated landscape by rising sea level. We come full circle, returning to the shallow sea where the story began.

Climate change was the cause of the glacial advances and retreats that generated the fluctuations in global sea level during both the Pennsylvanian and Permian. Although the continental glaciers were far to the south, and Kansas was at the time about 10 degrees north of the equator, the climate changes associated with the glacial and interglacial cycles are recorded by the rocks of the midcontinent. The consistent vertical pattern of climatically sensitive sedimentary features described above indicates that the climate changed from arid to progressively wetter and more seasonal conditions during the formation of each rise and fall of sea level. This climate pattern contrasts with that of the Pennsylvanian, which began with extensive lowland peat-forming swamp forests. A major long-term global change to increasingly drier and warmer conditions was occurring.

FLINT HILLS LANDSCAPES

The cyclic pattern of alternating terrestrial and marine sedimentary rocks characteristic of the Flint Hills can be seen in the faces of waterfalls. The resistant marine limestone units form the ledges at the tops of the waterfalls with the softer, colorful mudstones below. The colorful red-and-green banding of the mudstones represents stacked fossil soils that record multiple ancient vegetated land surfaces. In higher falls, or where a series of waterfalls is present, multiple cycles (or cyclothems) can be seen.

Prather Creek Falls displays the terrestrial mudstones of the Blue Rapids Shale that are overlain by the Funston Limestone. Geary County Lake Falls exposes the interval from below the Funston Limestone to the overlying Speiser Shale and the falls-capping Three Mile Limestone. The "barber pole" appearance of the Speiser Shale, resulting from the stacking of fossil soils, is

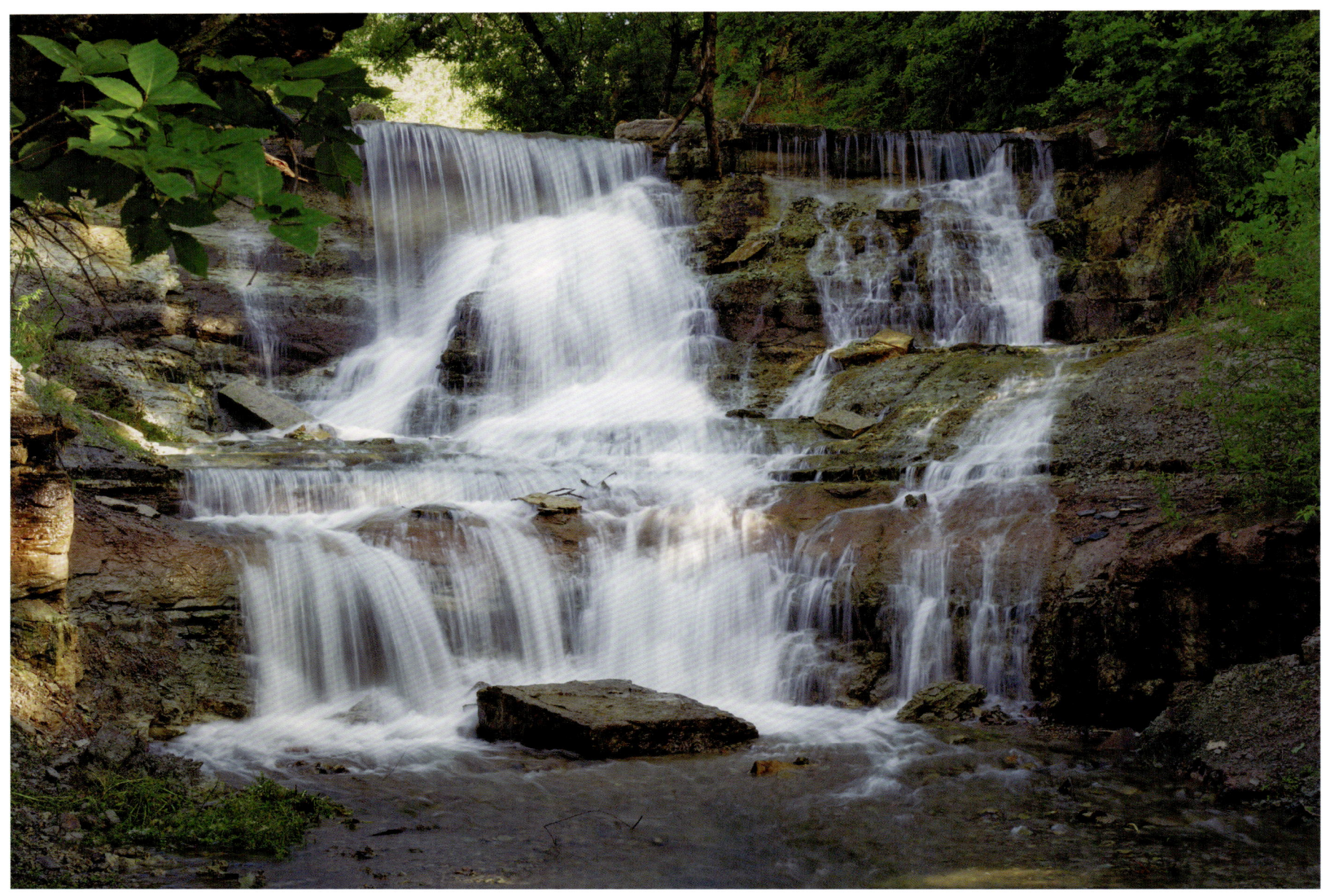

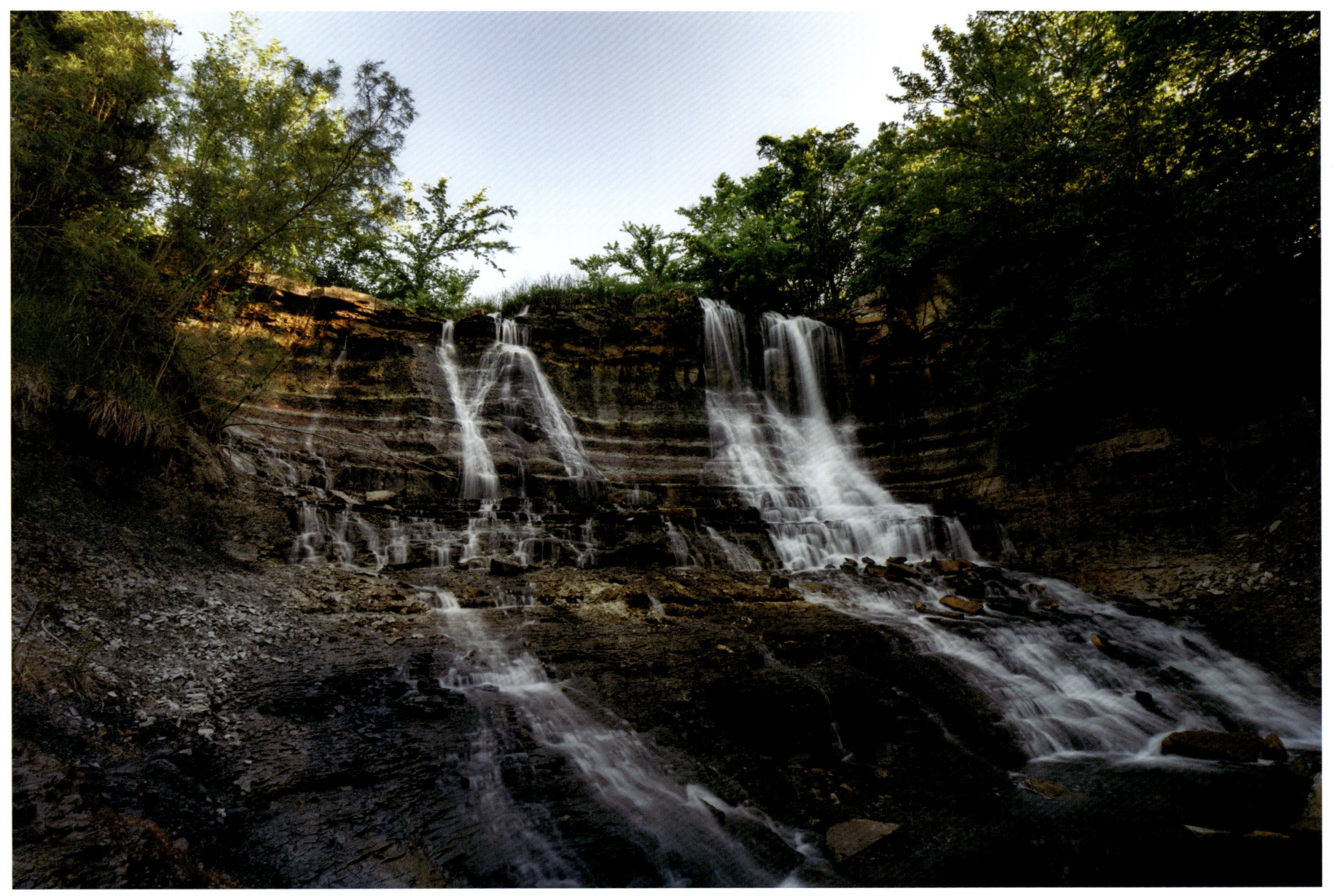

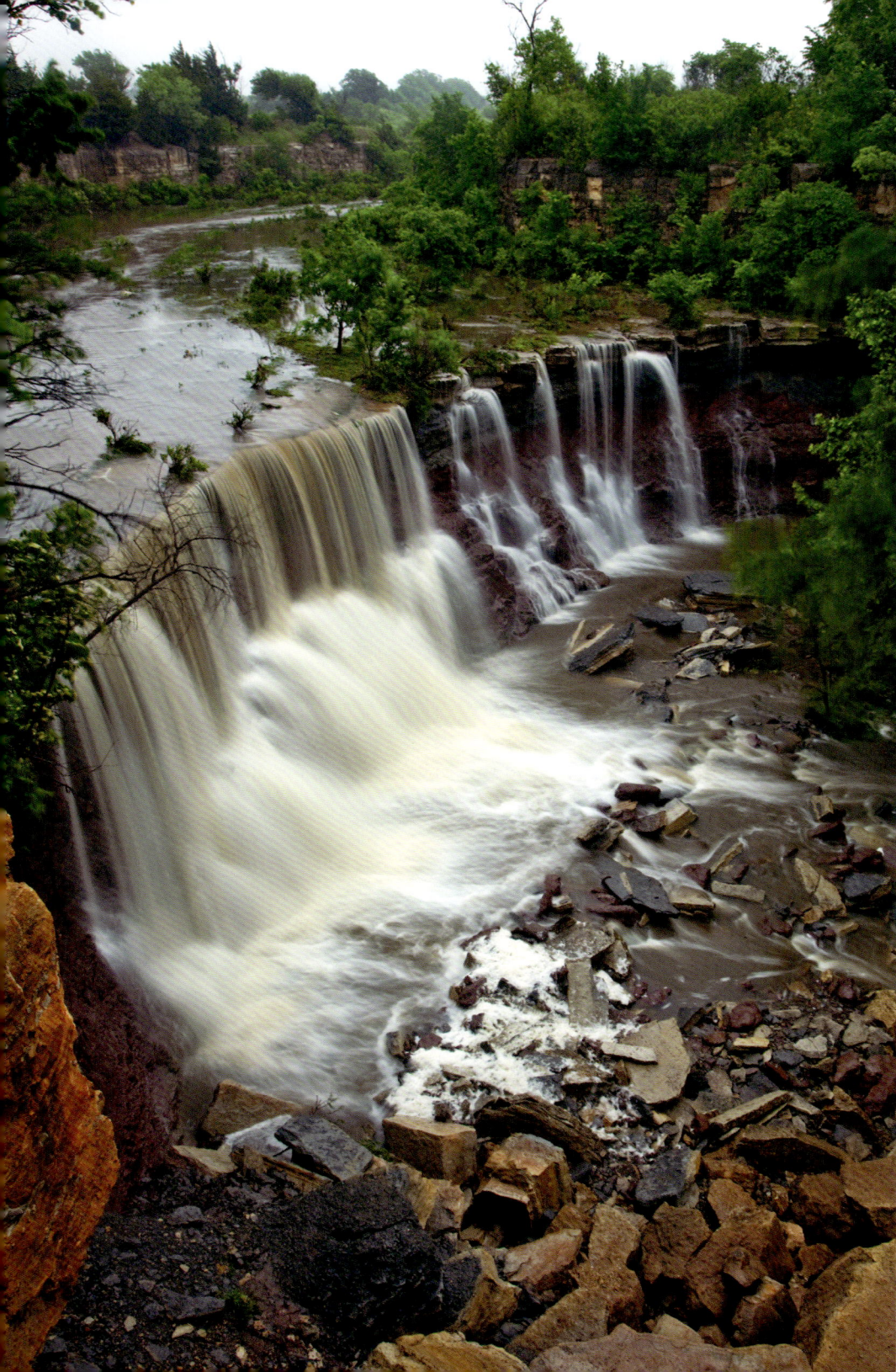

Page 30: The waterfalls where Prather Creek flows out of Chase County State Lake are popular places for people to enjoy the sights and sounds of the water cascading down three separate sets of falls. The middle set of falls, shown in this photo, is the largest and creates great photo opportunities when heavy rains have the falls flowing strongly. The use of a slow shutter speed, which causes the moving water to blur, renders the moving water a silky curtain of white in this photo.

Page 31: The waterfall at the outlet of the Geary County State Lake is the largest in Kansas. This wide-angle photo shows the span across the falls with a few different streams of water running over. This waterfall can have many different looks depending on how much water is flowing and can often be heard well before you see it when the water is flowing heavily.

Left: A classic view of the waterfall at Cowley State Fishing Lake. This is one of the most popular waterfalls in Kansas, and certainly one of the most beautiful. (Photo by Mickey Shannon.)

especially well-developed in this unit. Cowley Falls, near the Oklahoma state line in Cowley County, is at the southern end of the Flint Hills region. The thick Florence Limestone forms the cliffs upstream and caps the falls, below which are the red mottled mudstones of the Matfield Shale.

The photos below from Geary, Wabaunsee, and Greenwood Counties draw your attention to the rock-strewn crests of the hills and terraces of the Flint Hills. These weathered limestone rocks are what "hold up" these hills and form the characteristic terraced landscape. Many of these limestone layers also contain flint nodules that increase their resistance to weathering and erosion. As the limestone is slowly dissolved by the slight acidity of rain, the flint remains and accumulates at the surface over time. The flat terrace surfaces thus have thin soils and are covered in limestone and flint. The easily accessible flint was used by the native peoples of the Midwest for the manufacture of arrow and spear points and other tools.

The sloping areas below the limestones are underlain by the more erodible shales and mudstones. The deeper modern soils developed on these slopes are covered in prairie grasses and diverse flowers characteristic of the tallgrass prairie. These plant communities can be seen in areas of preserved native prairie such as the Konza Prairie (named after the local Kanza native people) and the Tallgrass Prairie National Preserve. In the valley bottoms, the deep, rich soils and available water support grasses, such as big bluestem, which can reach heights in excess of six feet with roots that extend just as deep. By contrast, the rocky, shallow soils of the terrace tops are water-stressed and have a different and more stunted plant community more characteristic of the shortgrass prairie to the west.

The wide-open vistas of the Flint Hills, where you can see into the next county (or the next state) and the horizon appears as a horizontal line, give the initial impression of a flat landscape. But this impression is deceptive. In reality, the landscape is deeply dissected by branching stream networks that have exposed a terraced (or stepped) topography resulting from the alternating limestones and mudstones of the underlying geology. The thickest and most flint-rich limestones form prominent flat terraces. For example, the Florence Limestone, which is the thickest limestone in the Flint Hills region, forms one of the most extensive of these terraces that create the appearance of a flat horizon. But even thin limestone layers a couple of feet thick produce inflections in the slopes of the hills that can be traced across the landscape. This pattern is so consistent that individual rock layers can be mapped from aerial photos.

Vegetation also responds to the differences in the underlying rock types. The tops of the rocky limestone terraces have more dry-adapted plants. The shales and mudstones below the limestones have deeper soils and support more typical tallgrass plant communities. However, these clay-rich rocks are also less permeable to water flow than the limestones. In the limestones, rainwater can percolate downward through cracks and fissures until it hits the shales below. The result is that water flows horizontally along the top surface of the shales at the base of the limestone layers until it emerges on the prairie slopes. Woody and water-loving plants tend to grow along these seeps and springs at the bases of limestones. Looking out across the landscape, you can often see horizontal lines of trees and scrub following the topography. The deep soils of the valley bottoms and the availability of water along the banks of streams support forests of oak, elm, cottonwood, hackberry, walnut, and hickory. These riparian zones are also home to a diverse community of birds.

Fire is a critical part of the tallgrass prairie ecosystem. Without fire, woody plant species would invade the prairie and transform it into dry scrubland. Prairie plants have evolved with fire and adapted to withstand its effects. In the past, fires probably occurred on average every three to five years, started by both lightning strikes and as controlled burns by Native Americans. Research at the Konza Prairie Biological Station in Riley County has shown that more frequent burning tends to decrease the diversity of plant and animal species in the prairie, and long intervals between fires allow the invasion of nonprairie woody plants.

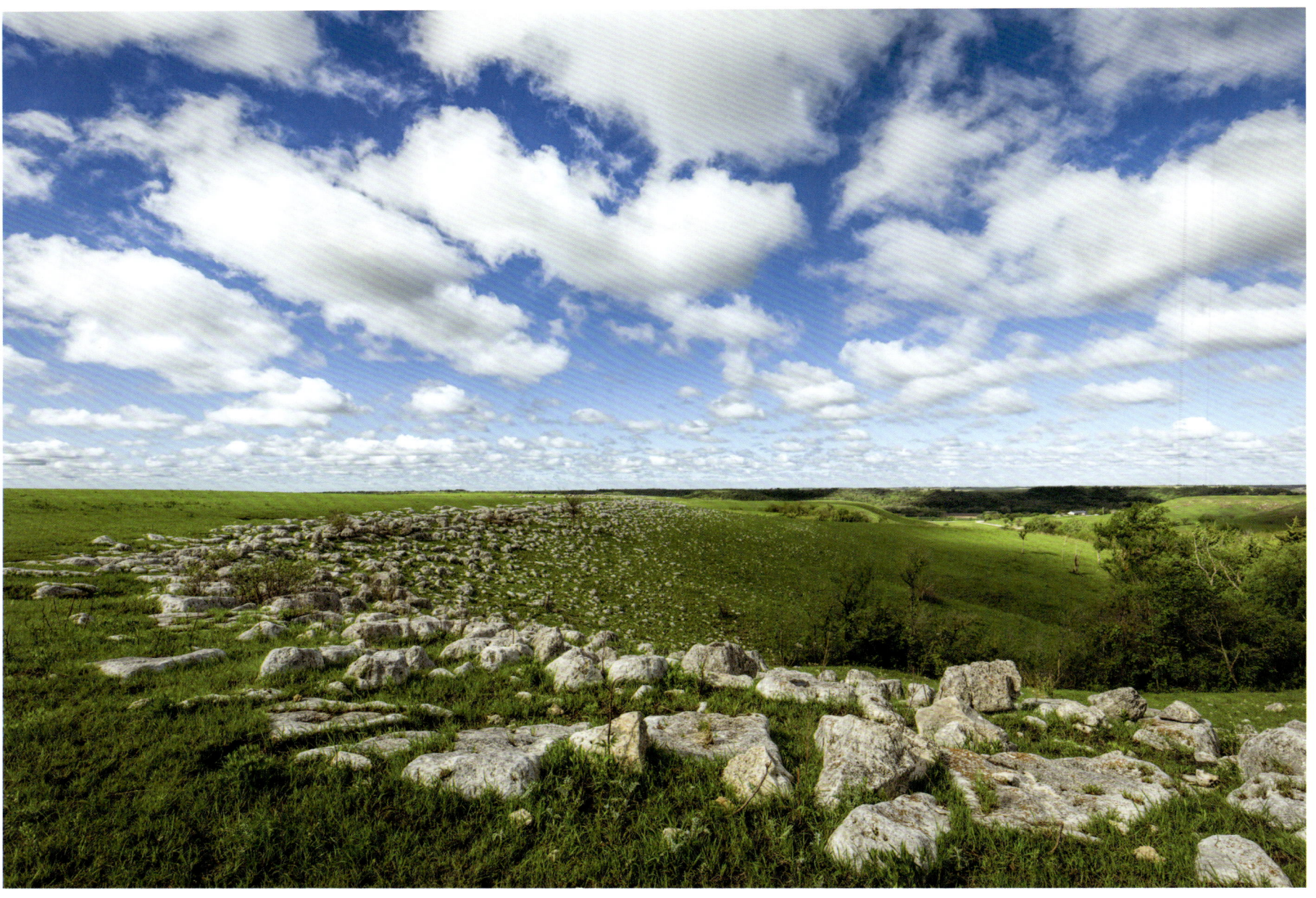

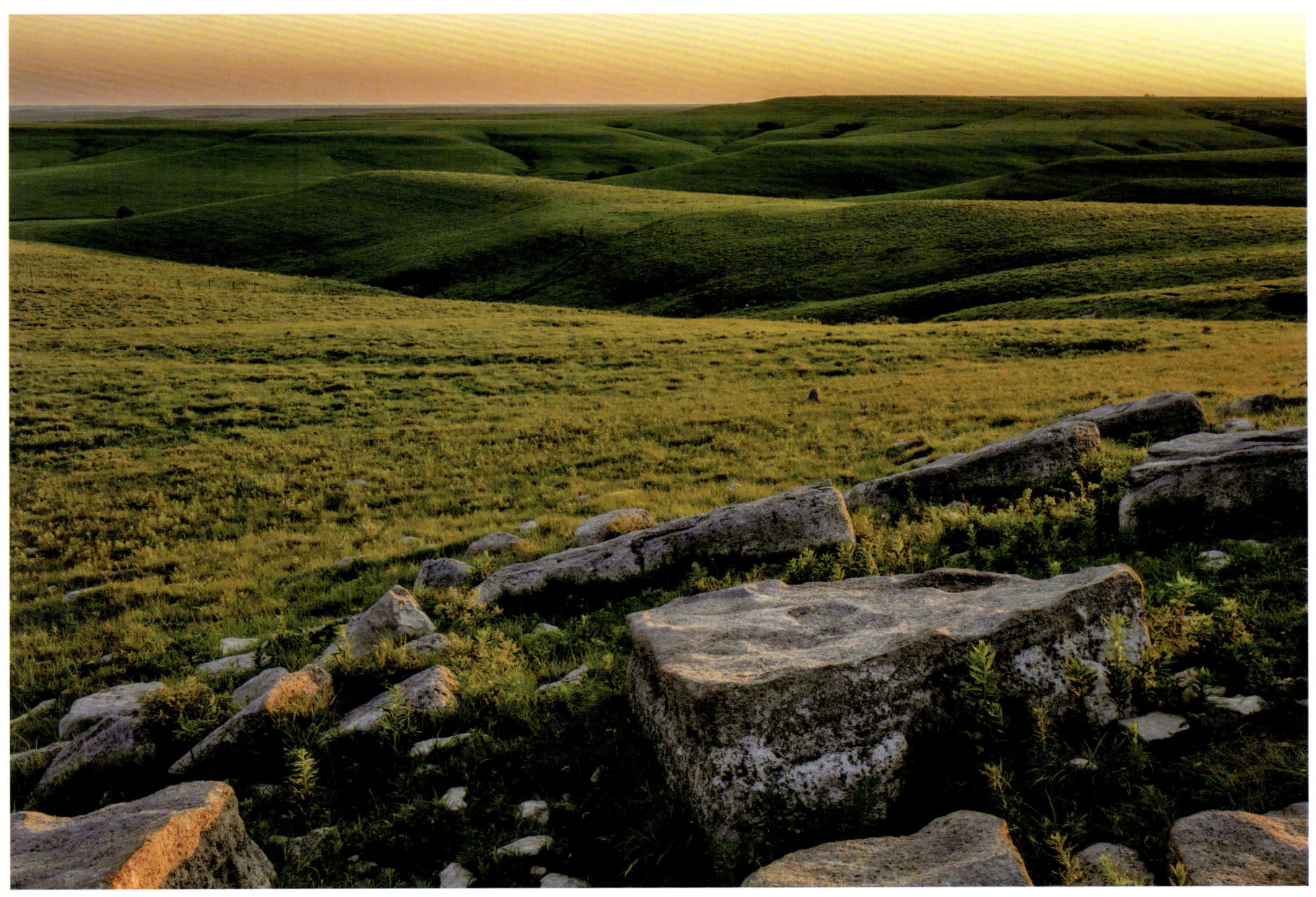

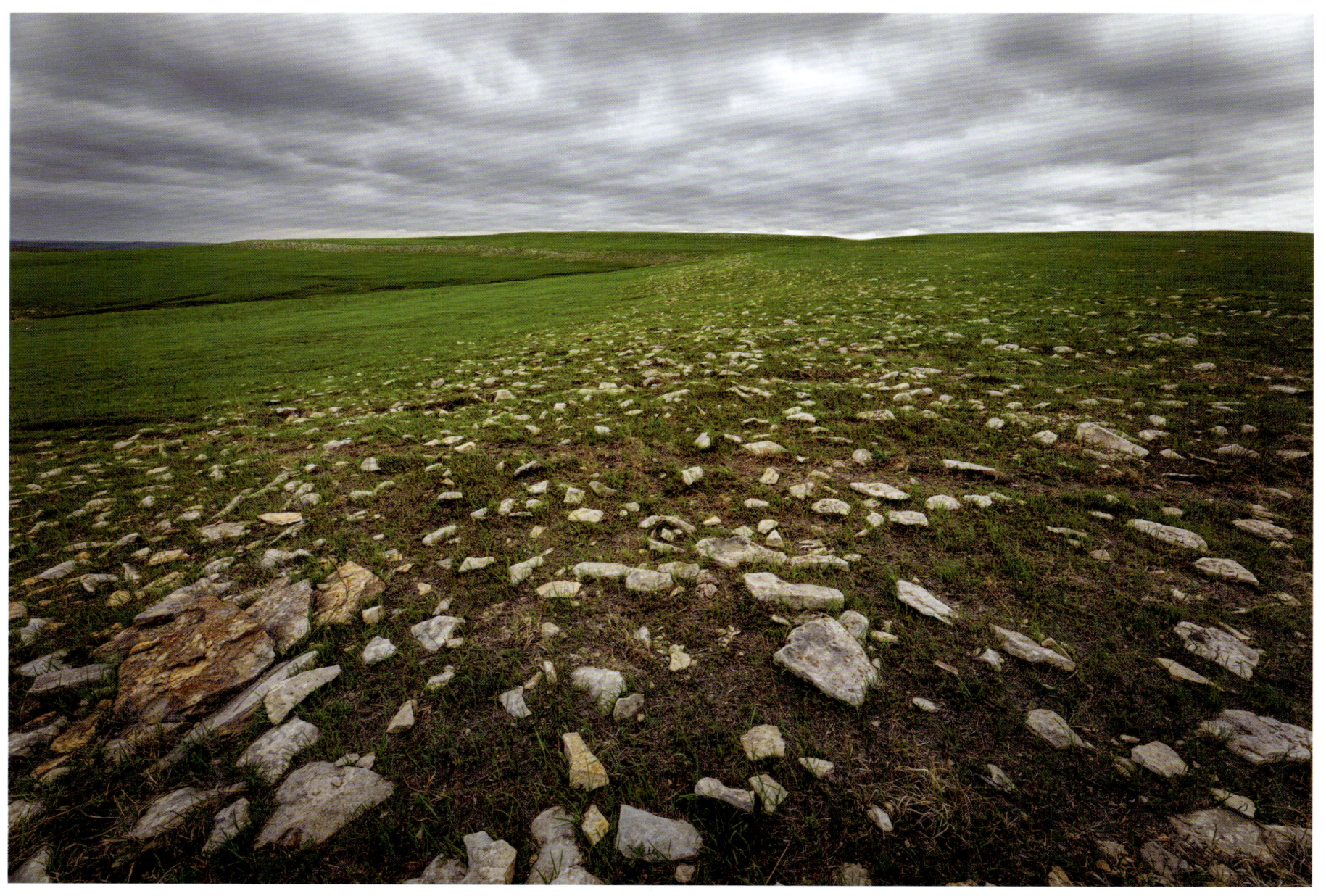

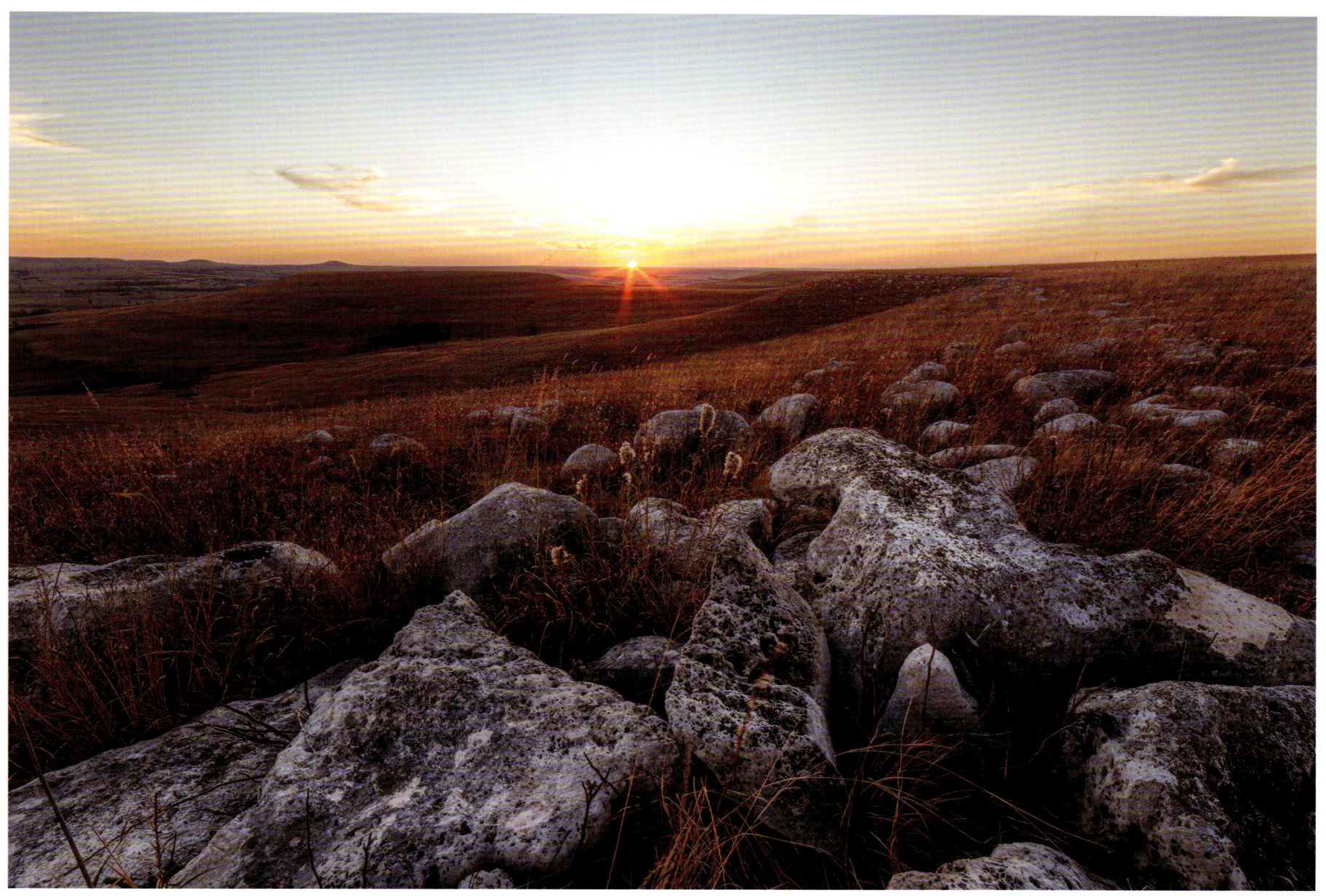

Page 34: The Flint Hills provide a wide range of photographic opportunities, and it is fun to wander county roads looking for interesting views to photograph. The rocks running around the edge of the hillside in Geary County caught my eye and formed a strong line in the photo, helping to draw the eye through the frame. The receding rocks also help create a sense of depth.

Page 35: Evening is a great time to photograph the Flint Hills because it wraps the textures and shapes of the landscape in soft light and color. The rocks in the foreground initially caught my eye in this scene from Greenwood County, and composing the photograph by placing my camera close to the rocks helped to bring out a sense of the distance from foreground to horizon. Being close to the rocks also serves to accentuate their texture, which contrasts with the softer texture of the prairie.

Page 36: Spring in the Flint Hills brings some welcome color back into the world after winter. The vibrant green of new growth after the controlled spring burns is delightful to see after the more muted browns of late winer. The rocks in this view of spring arriving on the prairie in Wabaunsee County lead the eye into an expanse of green.

Page 37: The rocks in the Flint Hills display a range of shapes and textures that provide a different sense of appeal for photographs and have always reminded me of the "bones of the prairie." In addition to the character the rocks provided, I included them in the foreground to help provide a sense of depth to the photo and bring out the vastness of the prairie here in Geary County.

Opposite: Early-evening light created a changing carpet of shadows and highlights and brought the shapes and textures of the Flint Hills to life in this view. Conditions like this are my favorite times to photograph the unique landscapes of the Flint Hills. The vibrant green of the prairie in early summer also adds to such scenes and shows how lush the prairie can be under the right conditions. This is one of my favorite views in Pottawatomie County.

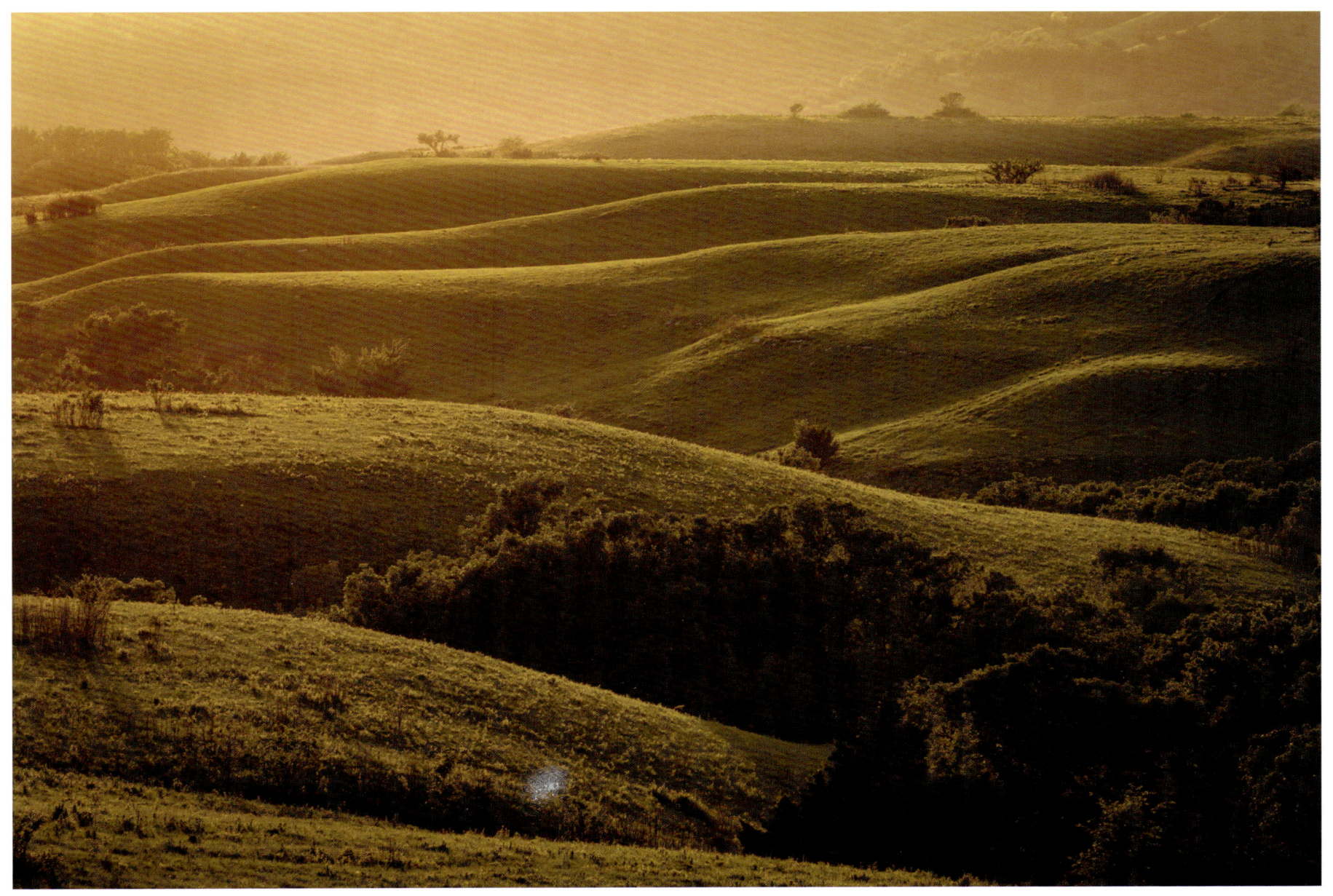

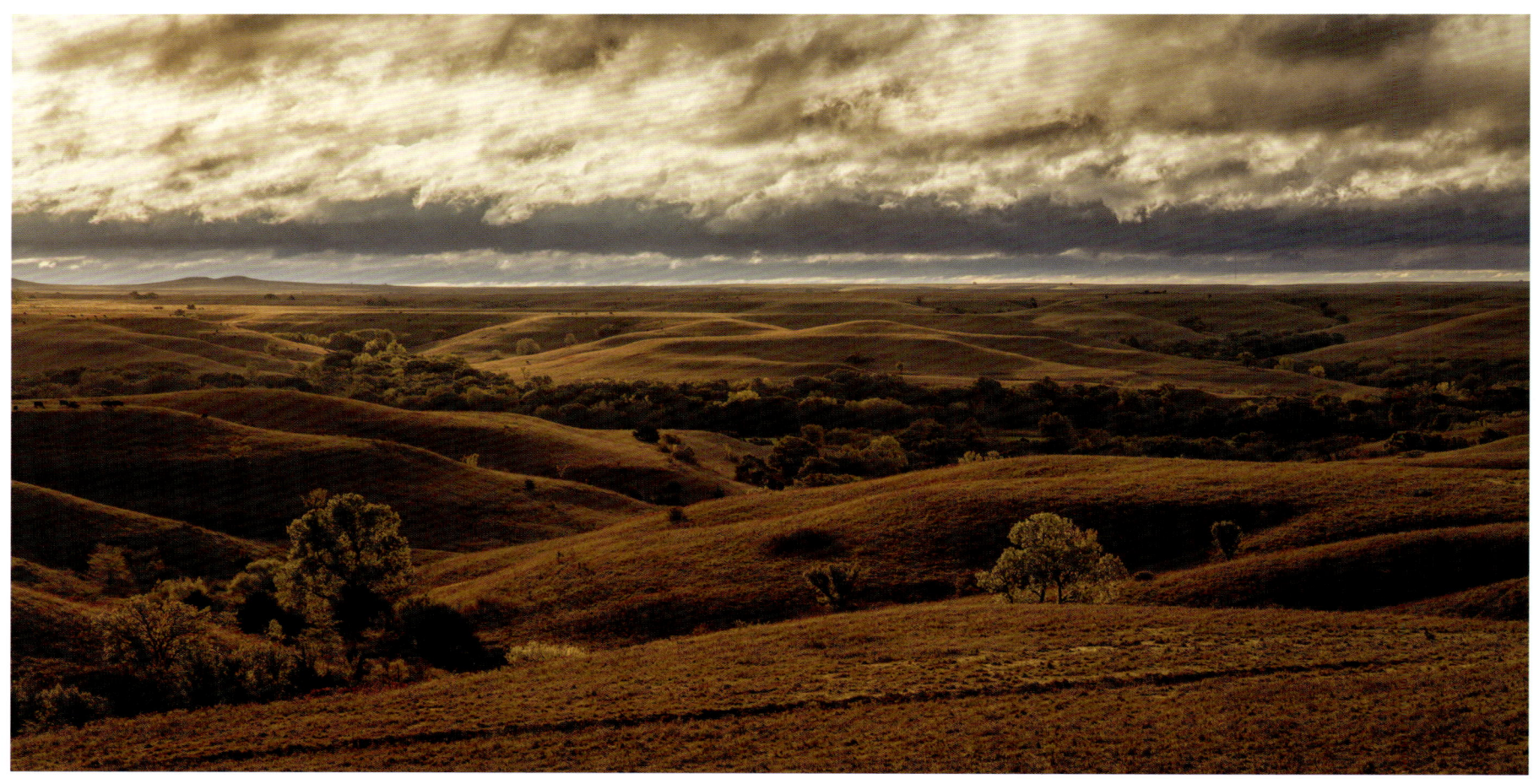

A cloudy fall morning in Pottawatomie County with sunlight breaking through the clouds made the Flint Hills look like they were composed of dunes in this view. I have always enjoyed watching the light and shadows chase each other across the hills and valleys of the Flint Hills, and that change in light constantly alters the texture of the landscape.

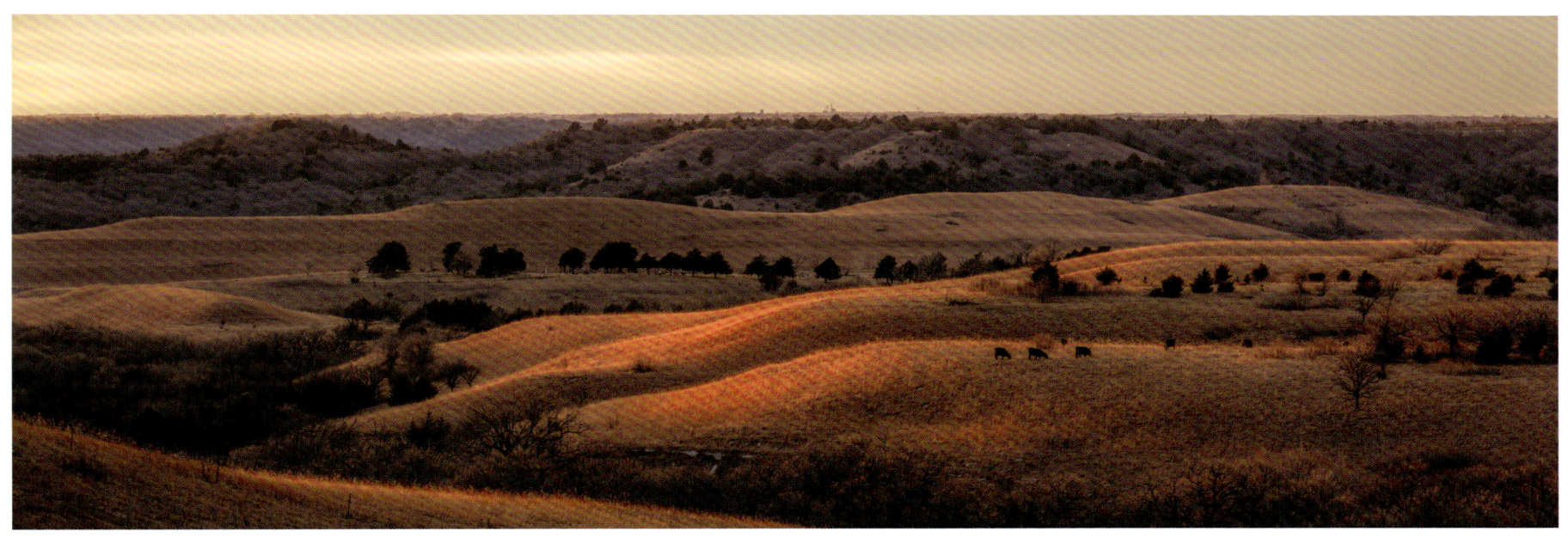

Spring usually gets the most attention for its color of the prairie in the Flint Hills, but the fall and winter prairie can also have beautiful colors and a lustrous appearance to the grass. When the light late in the day falls across the landscape, the shapes and textures stand out and grab your attention along with the colors of the season.

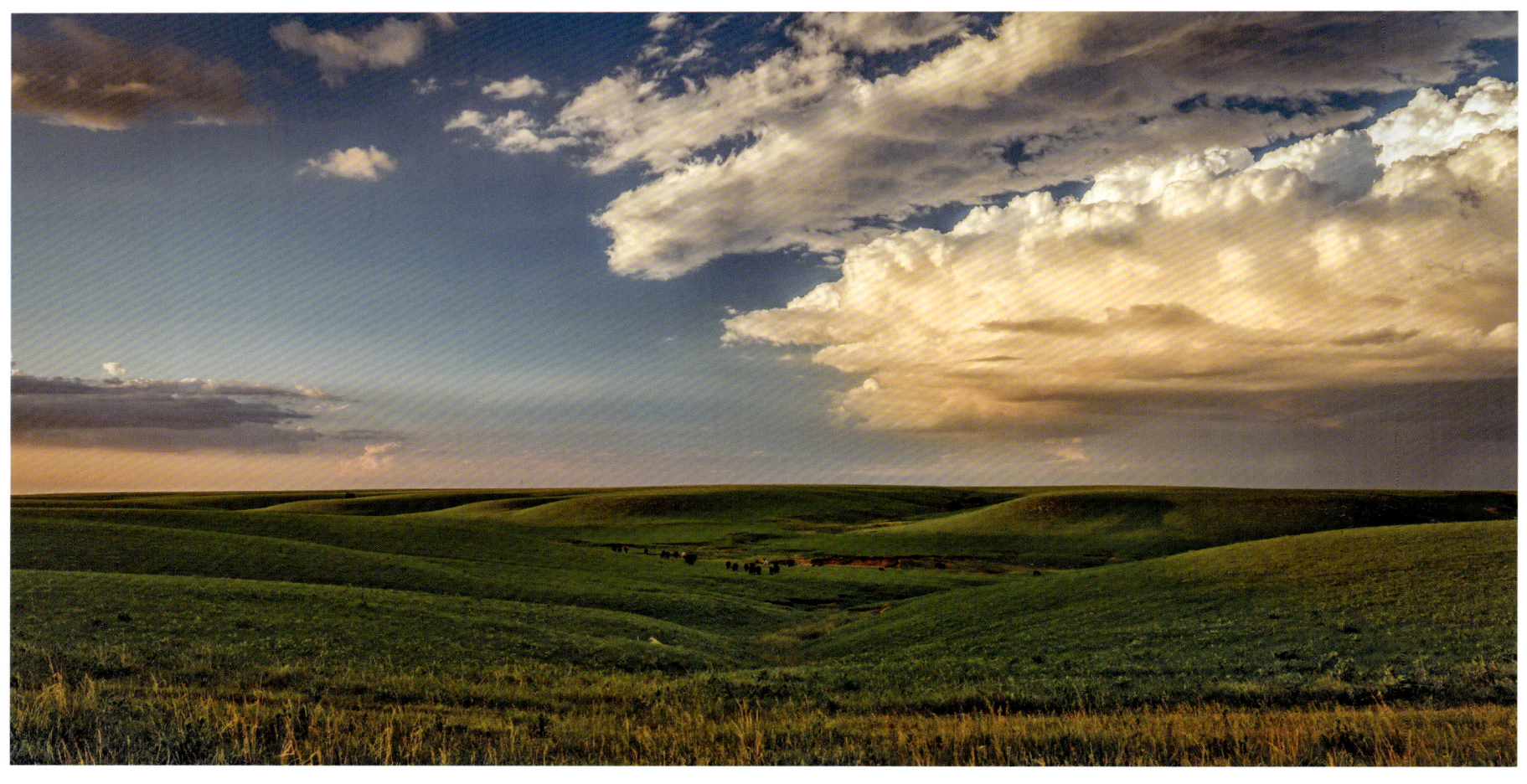

Above: The clouds building across the sky in this view from Wabaunsee County would eventually bring some spring rains to the area, almost always a welcome event. This scene had expansive views of both the prairie and the sky and highlighted the relationships between them. The vibrant greens of the new grass would not be possible without the storms and the rain that come through the area each spring.

Opposite: Looking down into the Kansas River valley south of Manhattan, the hills are framed by the golden sky and golden late-summer tallgrass. (Photo by Mark Clarke.)

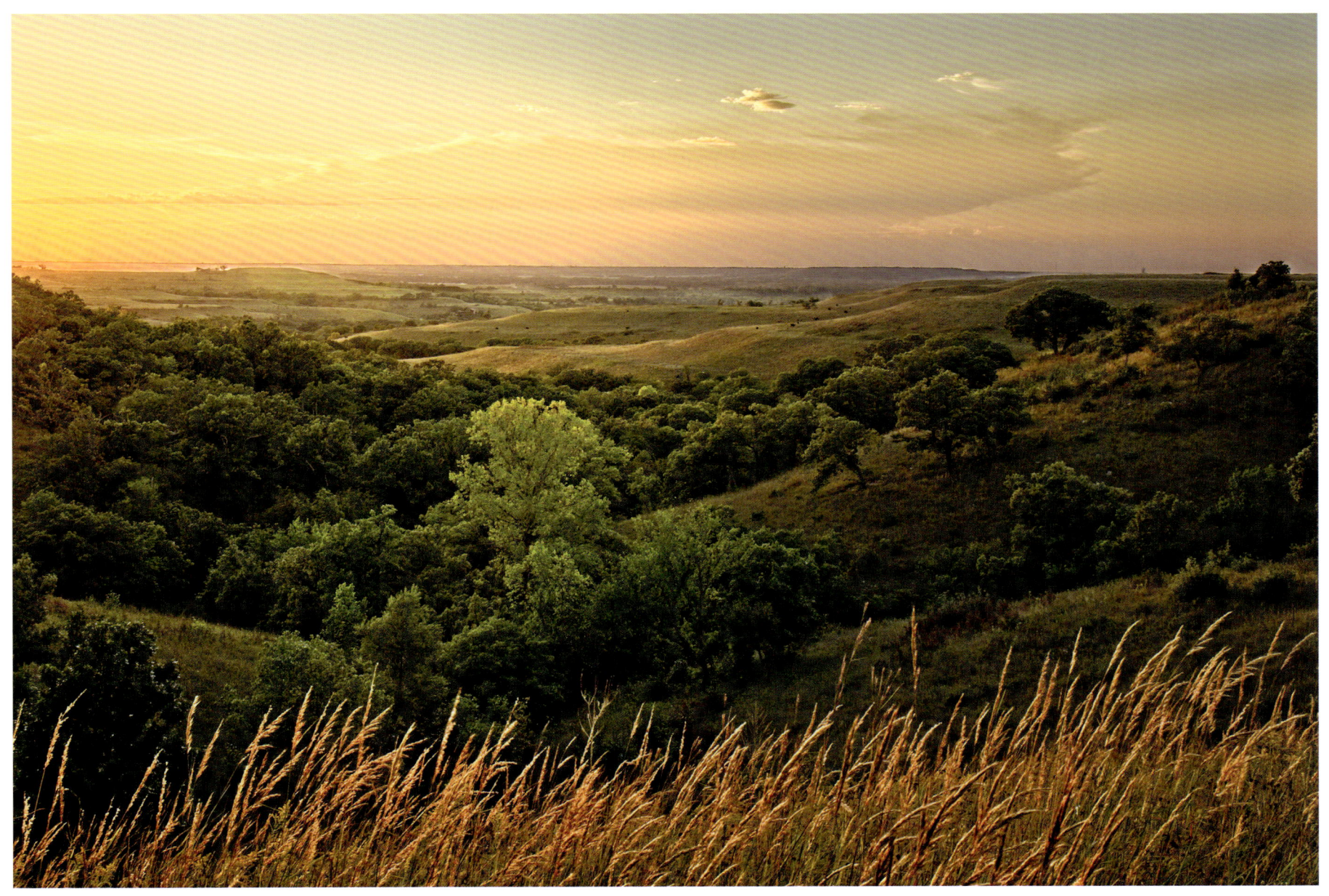

By removing standing dead grass and litter, fire increases the amount of water reaching the soil, recycles nutrients, and increases soil temperatures. For these reasons, fire stimulates the resprouting of grasses and the growth of soil microbes in the spring. Following fires, the prairie rapidly greens up with fresh growth and becomes very attractive to grazing mammals. Using this knowledge, native peoples burned the prairie to attract bison herds for hunting. Although this practice was discontinued by European colonists, controlled burning has been rediscovered and is widely used to manage grasslands.

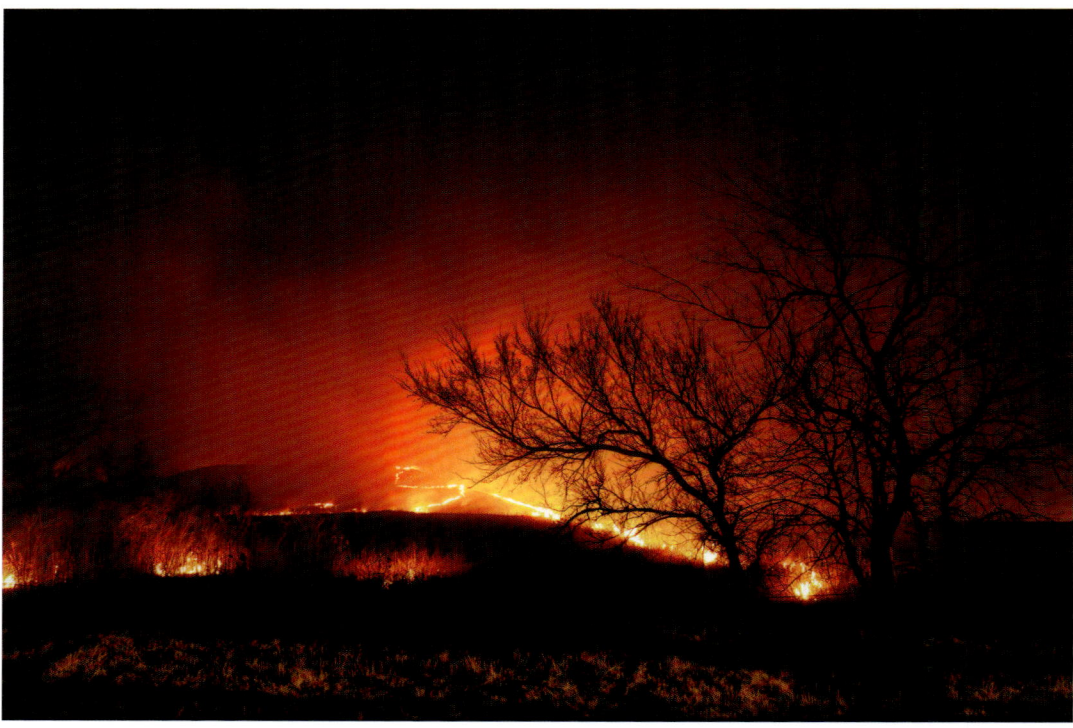

The controlled burns of the prairie in the Flint Hills are an important management tool for controlling trees and shrubs and maintaining the prairie. These burns are spectacular at any time, but at night, they are a unique experience. I was lucky to be able to photograph this spring burn in Riley County.

Opposite: Lines of fire during a controlled burn of the Flint Hills in Wabaunsee County.

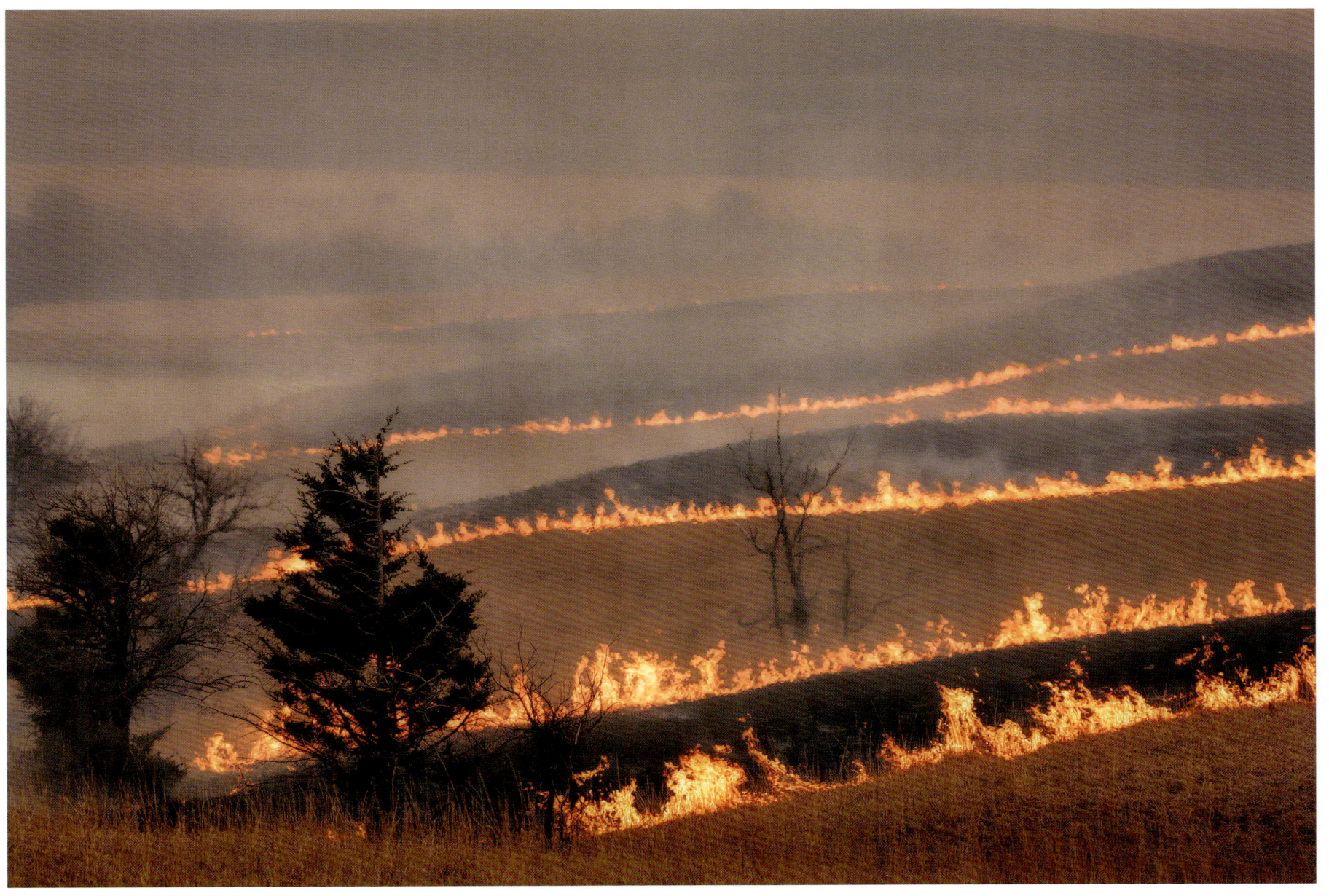

Section 4: Red Hills

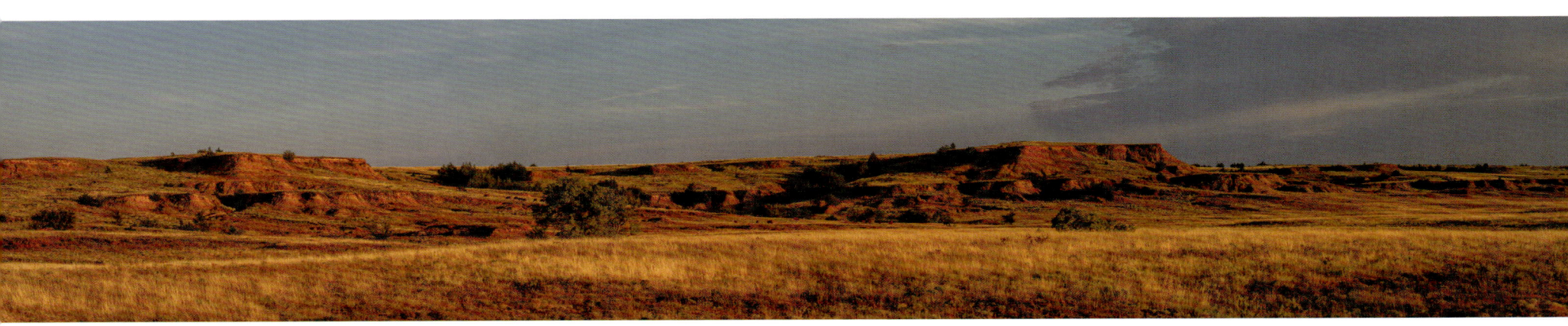

Rocks of the Red Hills and Wellington-McPherson Lowlands: A Drying Land

Geology of the Red Hills and Wellington Lowlands

The sedimentary rocks of the Red Hills and Wellington Lowlands record about 20 million years of the Late Permian from about 280 million years ago to 260 million years ago. The cyclical pattern of alternating marine and terrestrial rocks in the early Permian is replaced by rocks formed from sediments deposited in coastal settings and river floodplains.

The late Permian deposits of the Red Hills in southern Kansas include several thousand feet of red shales, siltstones, and sandstones deposited under dry climatic conditions. The red color is from the iron oxide commonly present in river-channel, floodplain, and fine windblown sediments. These red beds are interbedded with layers of gypsum, anhydrite, and dolomite formed by the evaporation of coastal waters. Surface weathering of these rocks has produced flat-topped red hills capped by resistant river-channel sandstones or by light-colored gypsum or dolomite layers.

The springs and streams in the area contain calcium and magnesium salts dissolved from the gypsum and dolomite beds. These salts were used as medicine by the native peoples of the plains, who called the area the "Medicine Hills." Sinkholes and caves, resulting from the solution of gypsum and do-

Previous page: This panoramic view in the Red Hills of Barber County captures a sense of space often found on the prairie, and the side lighting helped reveal the shapes of the distant rock formations. I grew up in the wide-open spaces of Kansas, and it isn't surprising that I find views like the ones shown in this photograph comforting. The Bean family settled in this area of Kansas in the late 1800s, and I imagine my ancestors looking out at this same landscape.

lomite, are also common features of the region. The collapse of the roofs of subsurface caves has produced the sinkholes as well as natural bridges.

In the Wellington-McPherson Lowlands farther to the north and east, the late Permian rocks underlie Pleistocene river deposits at the surface. This is an area of relatively flat to rolling topography. The western part of this region is underlain by one of the largest deposits of halite in the world, called the Hutchinson Salt. Gradual solution of the buried thick beds of the Hutchison Salt has resulted in the collapse of overlying rock and the subsidence of the surface. Lake Inman in McPherson County, the largest natural lake in Kansas, was formed by groundwater filling a depression caused by such subsidence.

A Story of Dramatic Environmental Change

Why the dramatic change in geology from the early to the late Permian? Two important environmental changes occurred over this period. The area that is now Kansas became isolated from the open ocean, and the climate became increasingly arid. Coastal marine environments were subjected to high rates of evaporation, resulting in brackish and hypersaline waters. These conditions resulted in the gypsum, anhydrite, and dolomite beds of the Red Hills and the thick halite beds of the Hutchinson Salt.

As the terrestrial environments became increasingly dry, eclian (windblown) silts and fine sands came to dominate the coastal plains, along with river-channel deposits. The types of fossil soils indicative of wet monsoonal conditions, found in the Flint Hills, are absent in the late Permian of the Red Hills.

The drying trend seen in the transition from the early to the late Permian continued into the subsequent Triassic and Jurassic Periods, which have little to no record in Kansas. This increase in aridity is recorded by extensive desert sand dune deposits in western North America, not unlike the great "sand seas" of the Sahara. The drying trend of the Permian was part of a global pattern of climate change.

The end of the Permian marked the largest extinction in Earth history. It was driven by a period of exceptionally high temperatures triggered by an influx of carbon dioxide from huge volcanic eruptions of thick lava flows in Siberia. These "flood basalts" constitute one of the largest volcanic events in Earth history, with an estimated volume of 750 thousand cubic miles and covering an area of about one million square miles. For comparison, the entire contiguous United States is about 3.1 million square miles in area. The high atmospheric carbon dioxide concentrations produced by this outpouring of basaltic lava led to a global runaway greenhouse climate, made worse by the release of methane from the melting of methane ices in the warming ocean floor. Global average temperatures increased by 6°C (11°F), with ocean surface temperatures in the tropics likely over 40°C (104°F) and those on land even higher. The vertical circulation of the oceans likely slowed to a halt, leading to very low-oxygen conditions over much of the ocean floor and the loss of bottom-living species. An estimated 95 percent of known marine species and 70 percent of terrestrial species became extinct as a result. This event also ushered in a dramatic change in the world's terrestrial life, including the rise of the dinosaurs, and of the first mammals, in the Triassic.

Red Hills Landscapes

The photos of Barber County wonderfully capture the rugged topography of the Red Hills region. The rust-red buttes and hills, with steep, angular slopes and flat grass-covered tops, are characteristic of the area. The area also has preserved native mixed-grass prairie, with both tall- and shortgrass species adapted to a semiarid climate. Brush and scrub trees colonize the valley bottoms along streams. This landscape is different from what many people would associate with Kansas.

Soft red siltstones and mudstones underlie the resistant sandstones that cap the flat-topped buttes and are easily eroded to form sculpted slopes. Although the differences in the rock types of the Red Hills are not visually pronounced, they do record a variety of ancient coastal plain environments. The beds or layers within these sediments record deposition on river floodplains, soils formed under dry climate conditions, and windblown sediments. All of these sediments reflect the increasingly arid conditions of the late Permian.

The subtle variations in the red coloration of the sedimentary rocks, combined with the absence of significant modern vegetation cover, create a very different landscape than that of the Flint Hills to the east and north. The superficially monochromatic landscape provides a dramatic contrast to the big open sky and distant horizons.

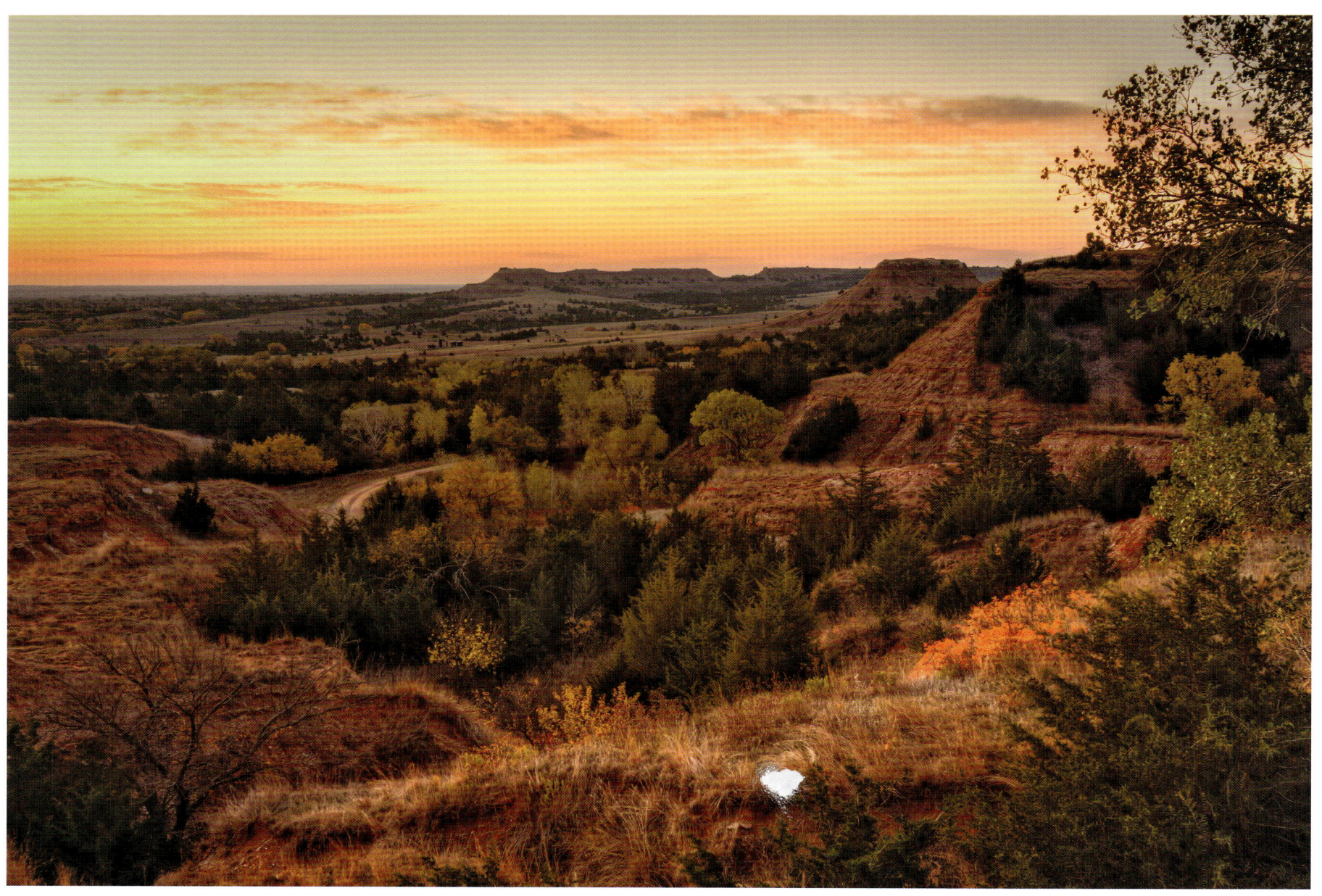

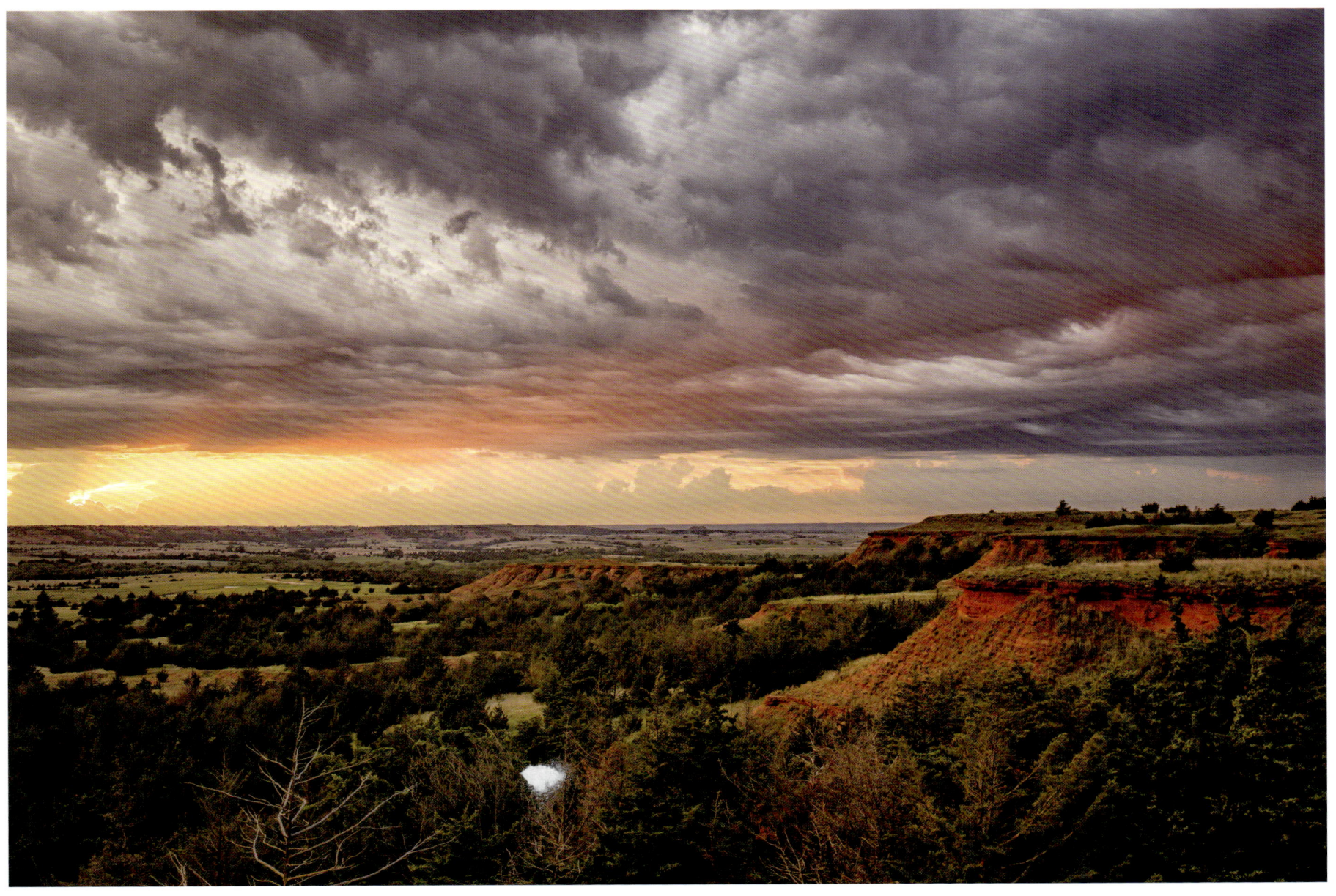

Page 50: This sunrise was the perfect time to photograph the Red Hills in Barber County. Besides the striking color, the view had a depth to it that I liked. The hint of the road in the center added to the sense of depth and visually drew my eye into the image.

Page 51: A storm was moving through the Red Hills in Barber County on this particular evening and created sunset colors that matched the colors of the hillsides. Storm systems passing through often create dramatic conditions and impart a special mood to a photograph.

This sunny day provided a brilliant blue sky that set off the color of this hill in Barber County. The color contrast between the sky and land can be very striking and provides a way to visually highlight the landscape in this part of Kansas.

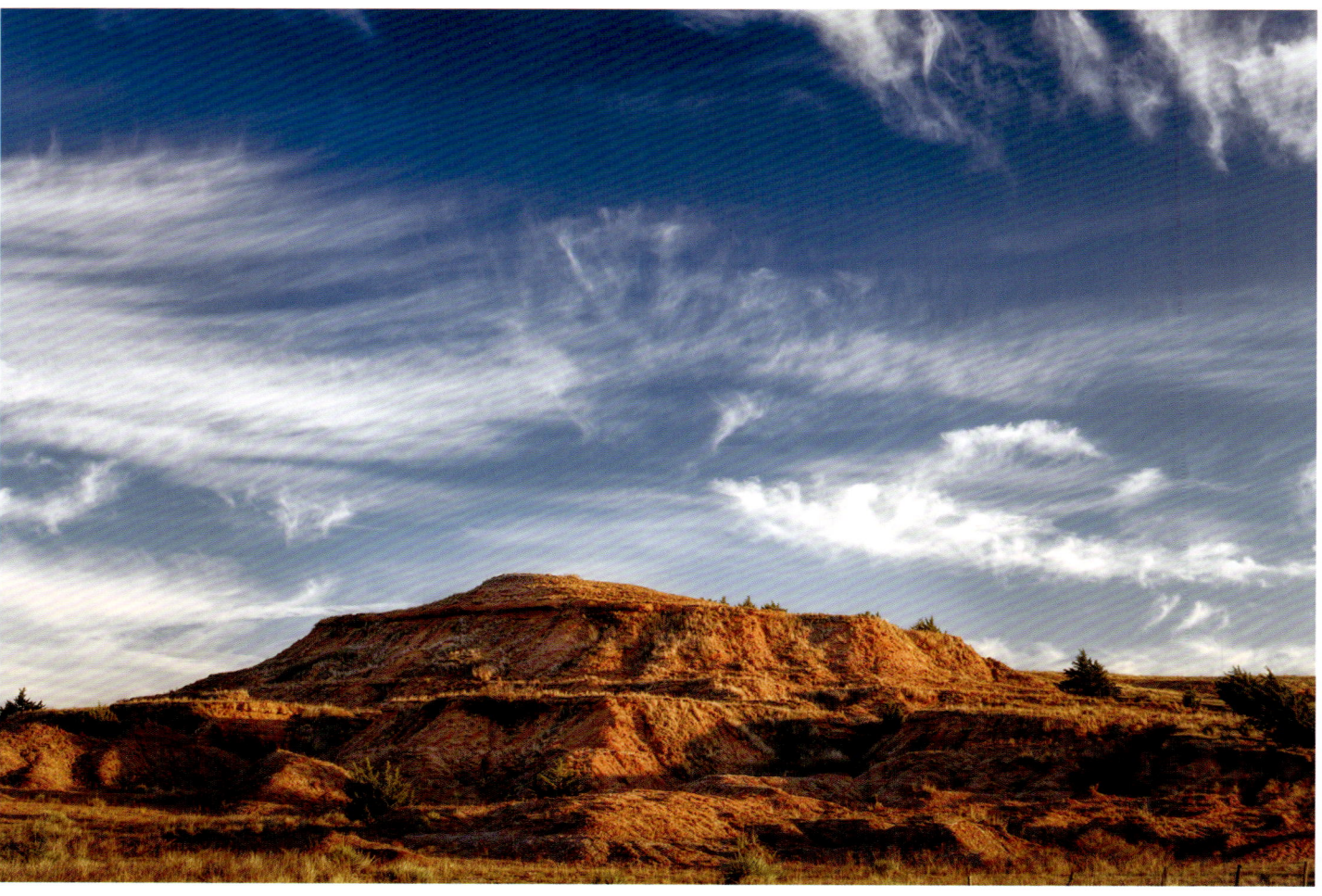

The more weather-resistant rocks that cap the flat-topped buttes and hills include river-deposited sandstones as well as gypsum and dolomite layers. The light-colored bands of gypsum and dolomite, as seen in the Medicine Lodge photo, have given the name Gypsum Hills to this area. Gypsum and dolomite are minerals formed by the evaporation of saline lakes or hypersaline coastal marine waters in the arid climate of the Late Permian. Although water-soluble,

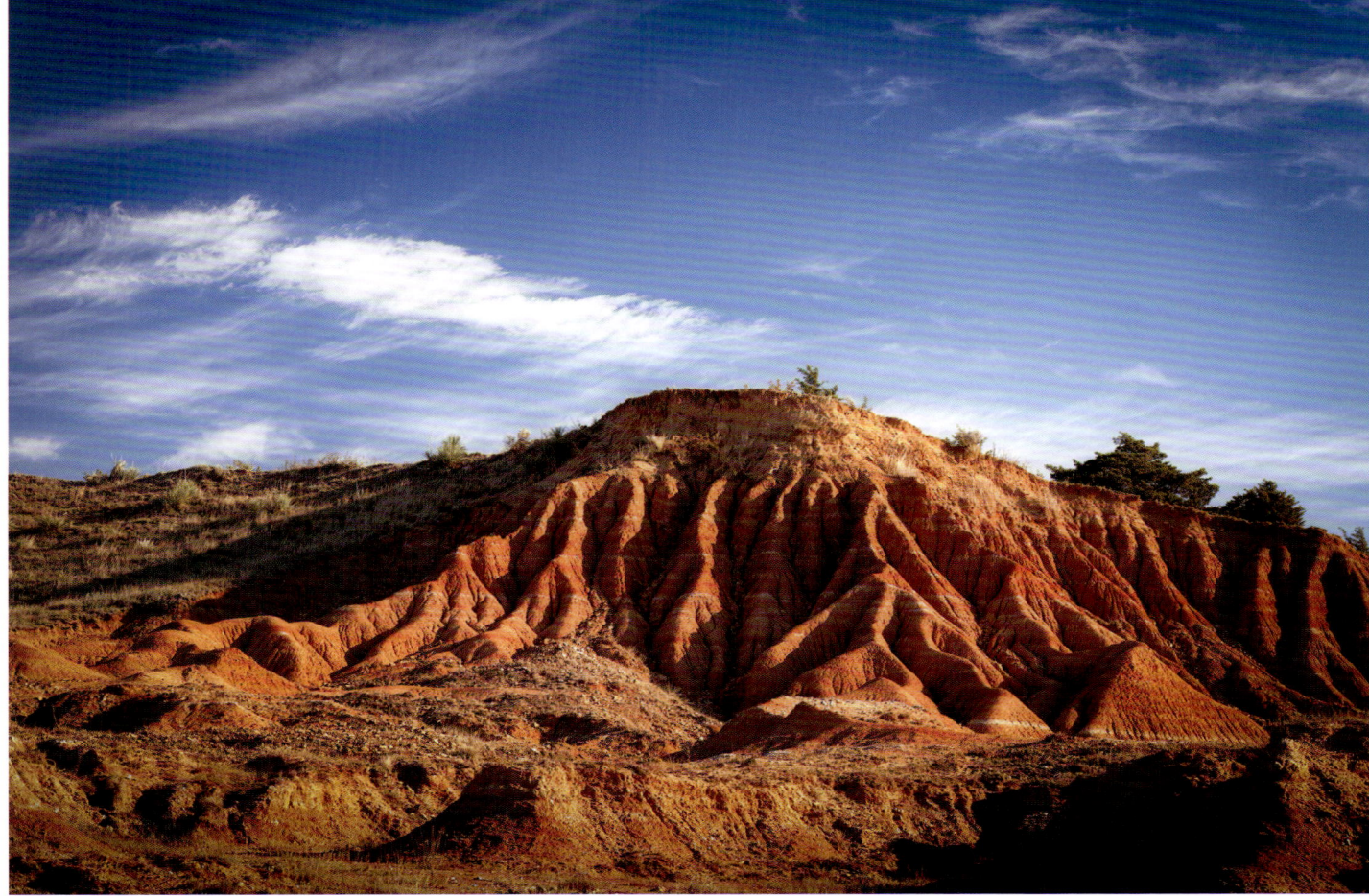

these gypsum and carbonate deposits are resistant to wind erosion and thus form solid capping layers in the modern low-rainfall environment of western Kansas.

Life in the late Permian would have been significantly heat- and water-stressed. This environment provided the conditions for the domination of more dry-adapted plants, such as gymnosperms (evergreen cone-bearing plants), and for the evolution of the reptile ancestors of both mammals and dinosaurs. An interesting coincidence is that the climate in this area today is not very unlike what it was in the late Permian over large areas of the low-elevation continental interior.

The angular topography of red buttes and narrow eroded draws of the Red Hills contrasts sharply with the surrounding mixed-grass prairie. These grass-

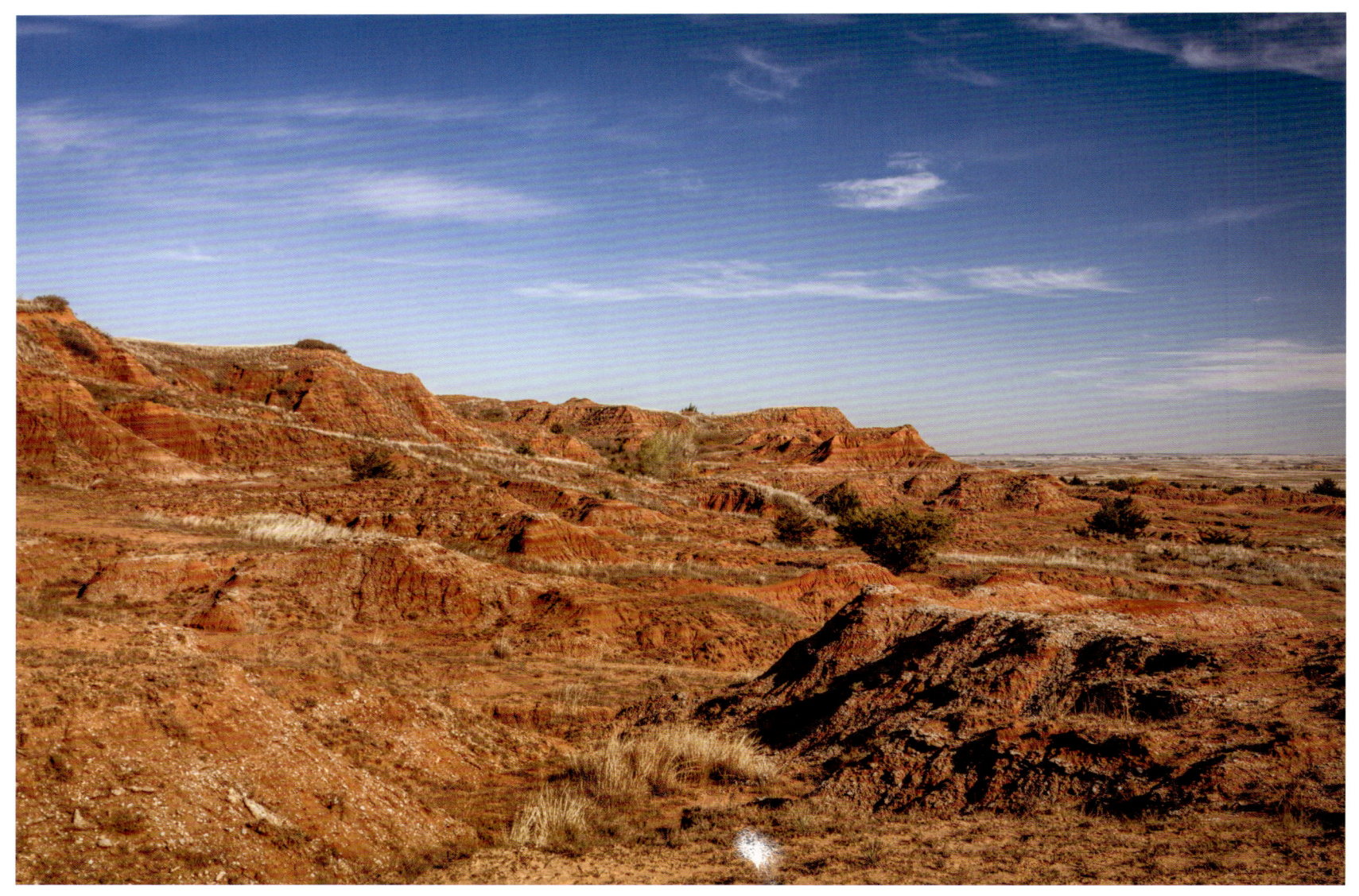

The view across this badlands area in Barber County is not typically what people expect from Kansas landscapes. This is another place that I feel I could spend a long time exploring.

lands, located in the dry subhumid to semiarid central part of Kansas, are a transition between the shortgrass prairie to the west and the tallgrass prairie to the east in the Flint Hills region. The Red Hills are at the southern edge of this ecological zone. Depending on soils, topography, grazing, and fire, the mixed-grass prairie can favor the tallgrass species of the east or the shortgrass species of the west. Woodlands and scrublands are restricted to lowlands near streams or to areas protected from fire.

The view down this draw, lined with trees starting to show their fall colors, caught my attention. The white outcroppings of rock running across the hill on the right side of the frame added some additional color to this view of the Gypsum Hills near Medicine Lodge in Barber County.

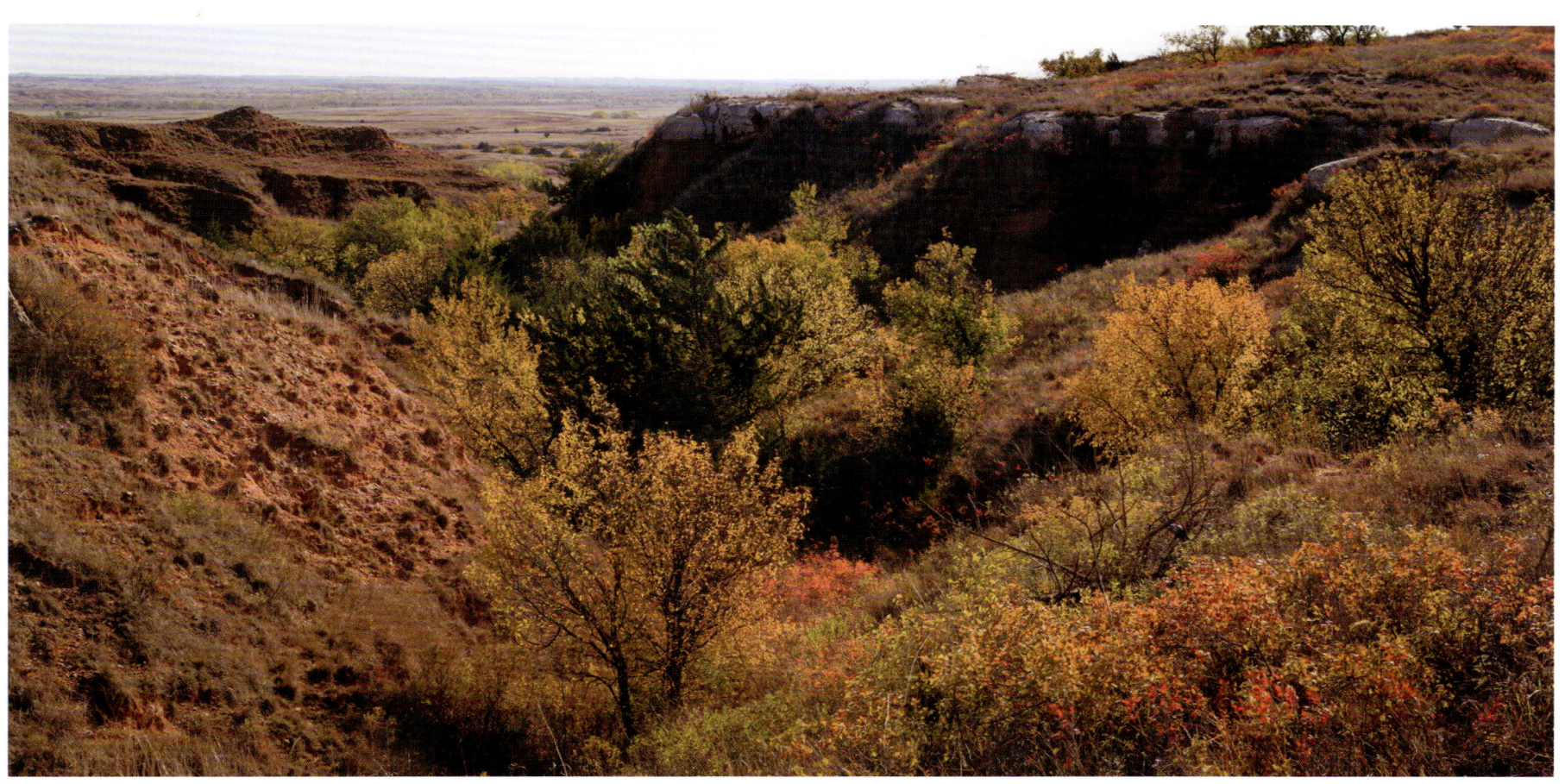

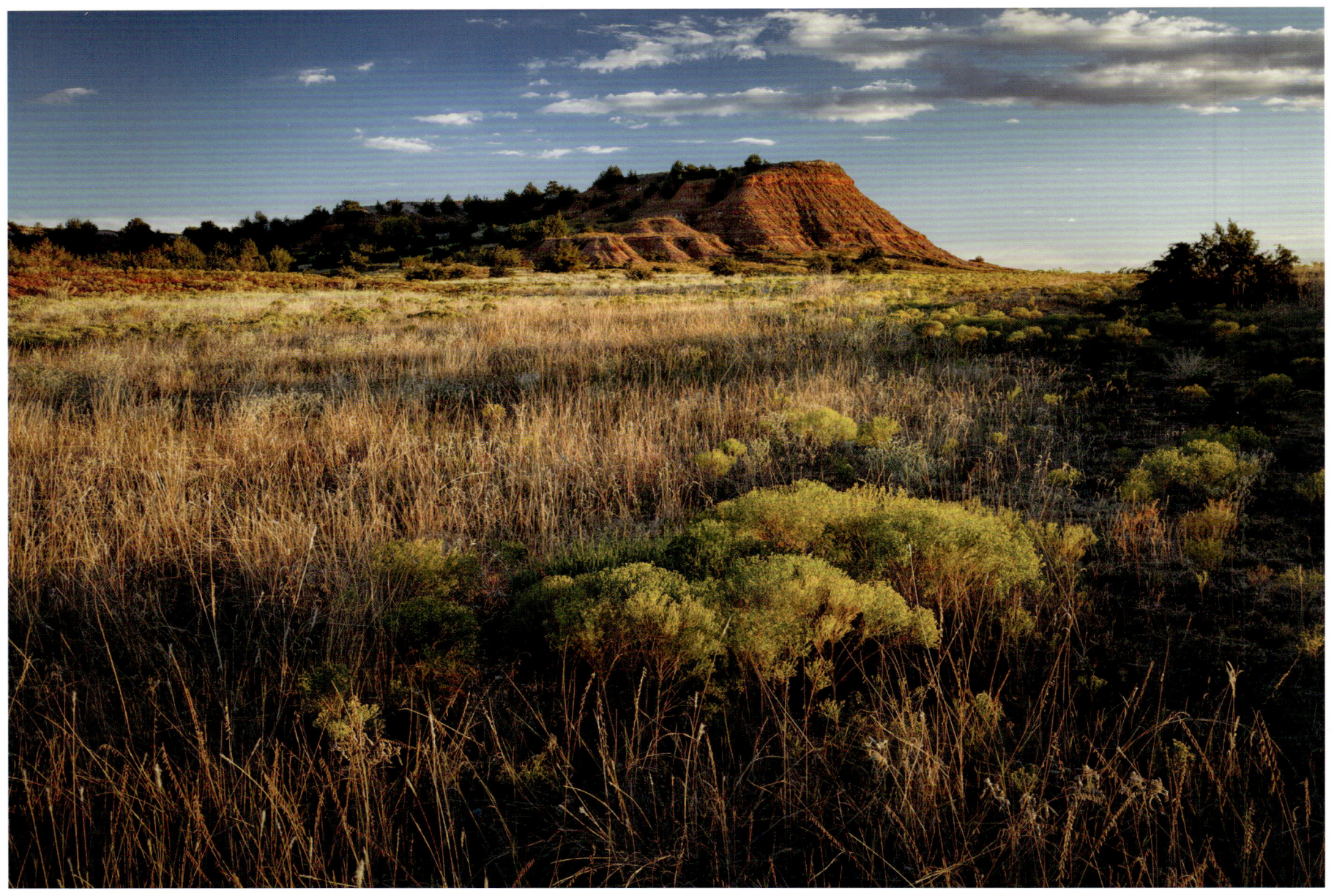

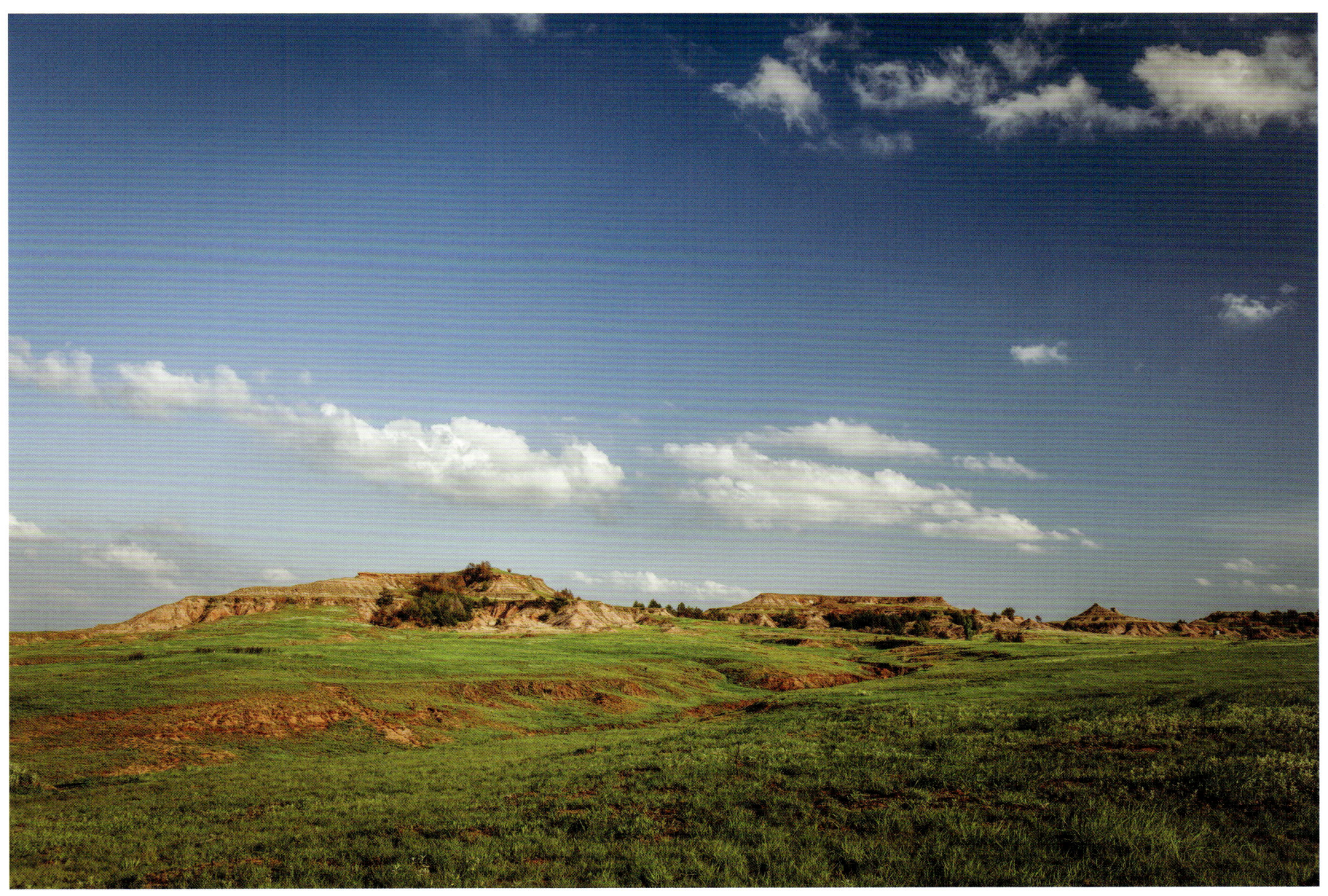

Page 56: I liked the way the shape of the broomweed flowers in the foreground mimicked the hill in the background in this image from Barber County.

Page 57: This colorful spring day in the Red Hills of Barber County provided an opportunity to showcase this region of Kansas in a year when the prairie was covered in vibrant and lush green.

Opposite: I've always enjoyed my visits to the Big Basin Prairie Preserve in Clark County. The quiet and the sense of perspective here invite you to slow down and be more present.

The Big Basin in Clark County lies within the Big Basin Prairie Preserve. The over 1,800 acres of native mixed-grass prairie in the preserve are managed by the Kansas Department of Wildlife and Parks. The area consists of rolling hills with level uplands and small canyons. The Big Basin itself is a large circular depression, about a mile in diameter and 125 to 150 feet deep, with a relatively flat bottom and surrounded by steep walls. It was formed by the solution of subsurface gypsum by groundwater and the subsequent collapse of the overlying rock into the void. Gypsum caves and sinkholes formed by the same solution-and-collapse process occur throughout the Red Hills.

St. Jacob's Well lies in a nearby small solution-collapse depression called the Little Basin. The well contains a small permanent pond surrounded by cottonwood trees. The spring-fed water fills a deep sinkhole, which reportedly has never gone dry.

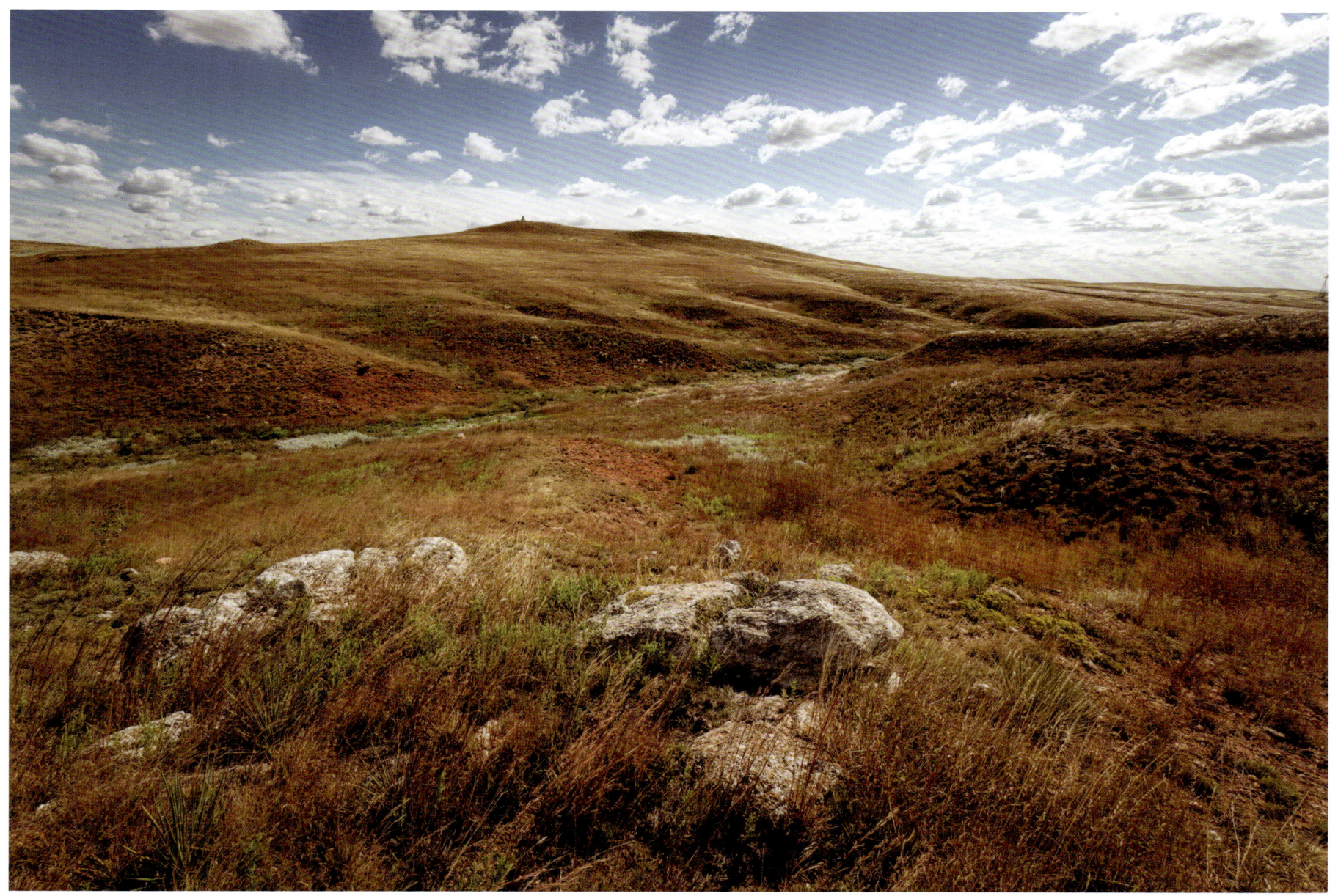

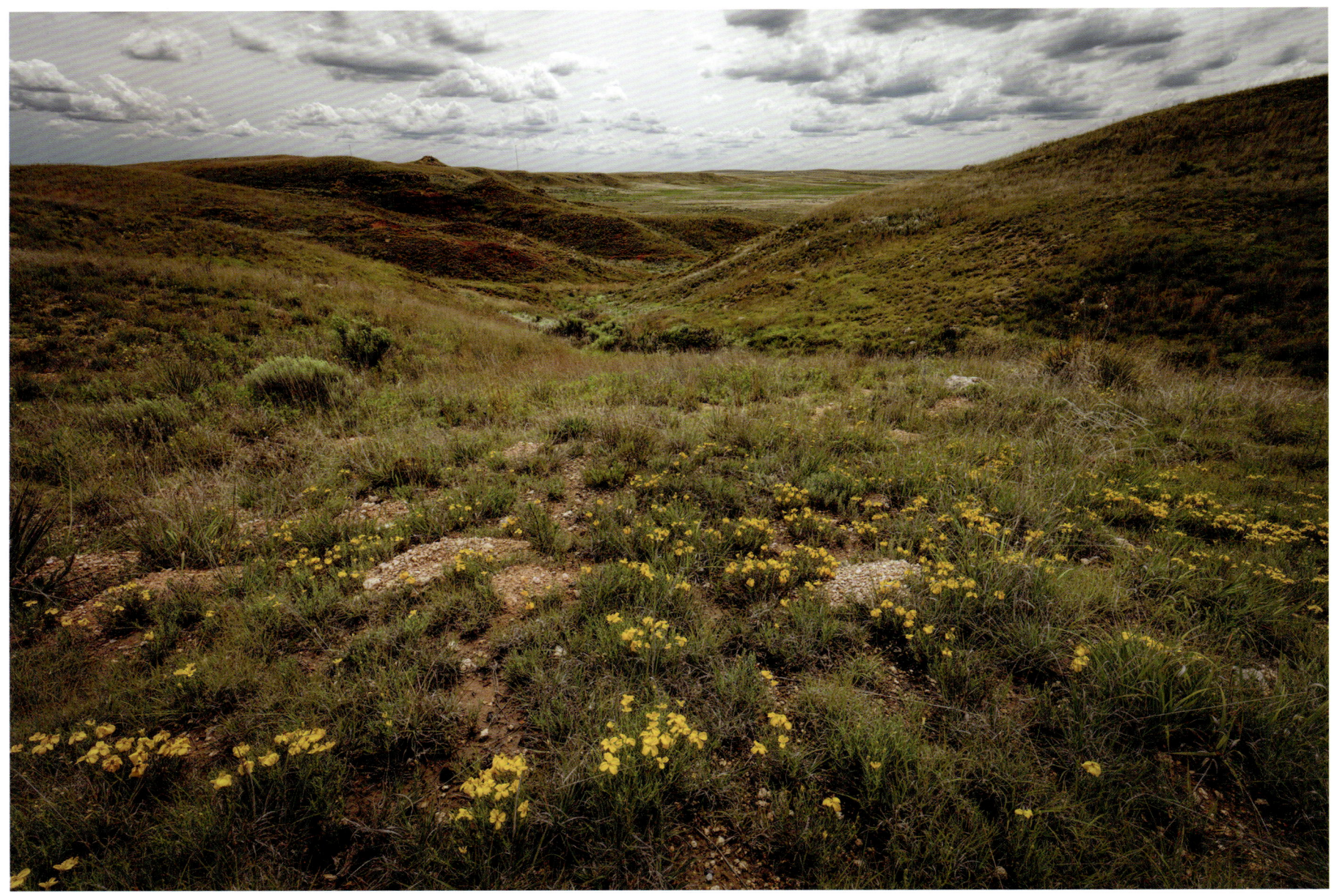

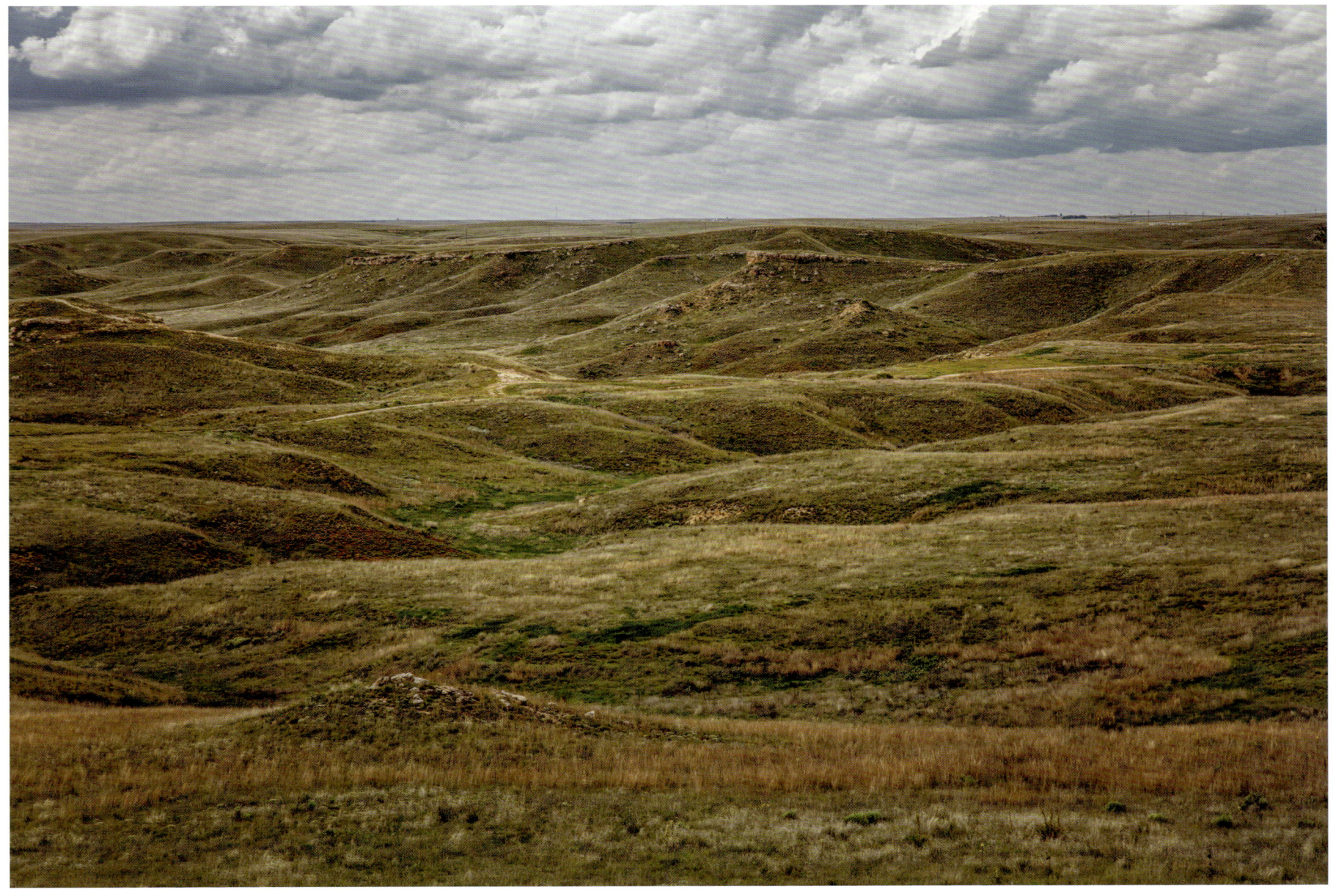

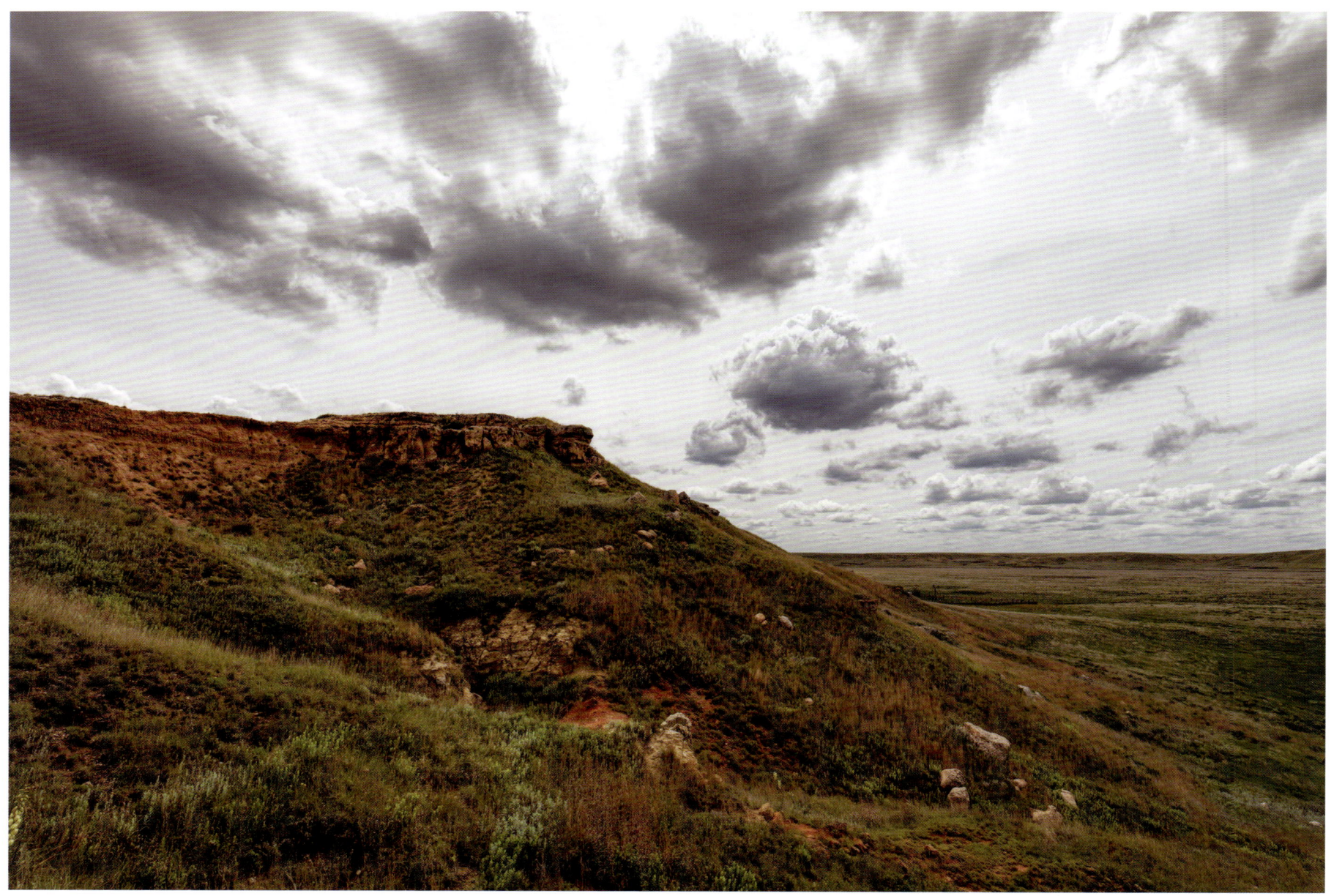

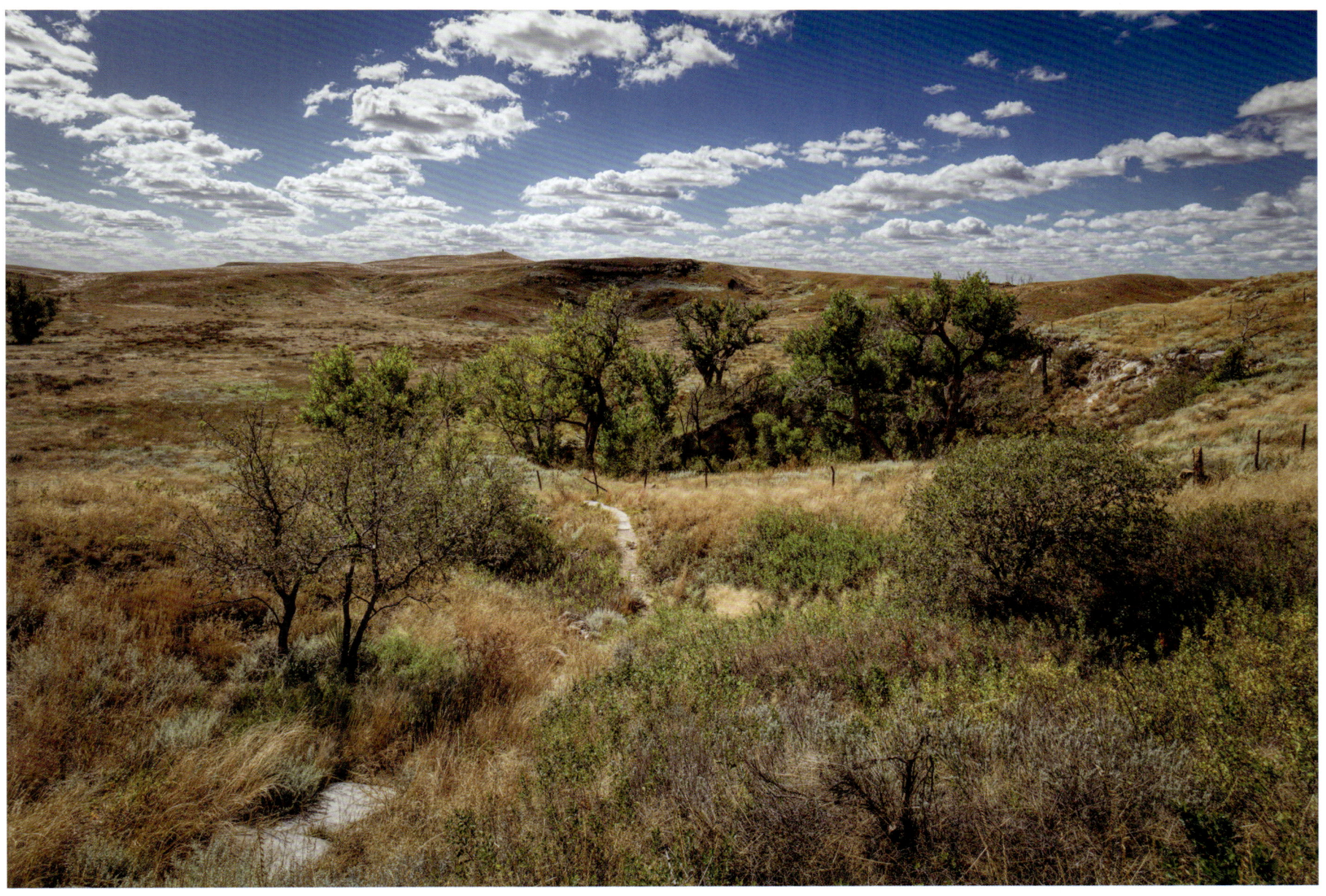

Page 60: The unique features of the landscape at Big Basin Preserve in Clark County are enough reason to visit, but being there when the wildflowers are blooming in the spring is a treat and creates new photographic opportunities.

Page 61: As a photographer, I'm drawn to the shapes and textures of landscapes and to the light that helps to reveal those features. I'm learning to ponder all the forces and effects of time that have gone into shaping a landscape as well, and this view from Big Basin in Clark County is a good place for that.

Page 62: Here we see the steep rim wall of the Big Basin dropping down to the nearly flat bottom of the basin to the right.

Page 63: This photo shows the path leading down to St. Jacob's Well in Clark County on a fall day. This site is interesting not only for its historical significance but also for the myths that have been reported about it over the years.

Section 5: Smoky Hills

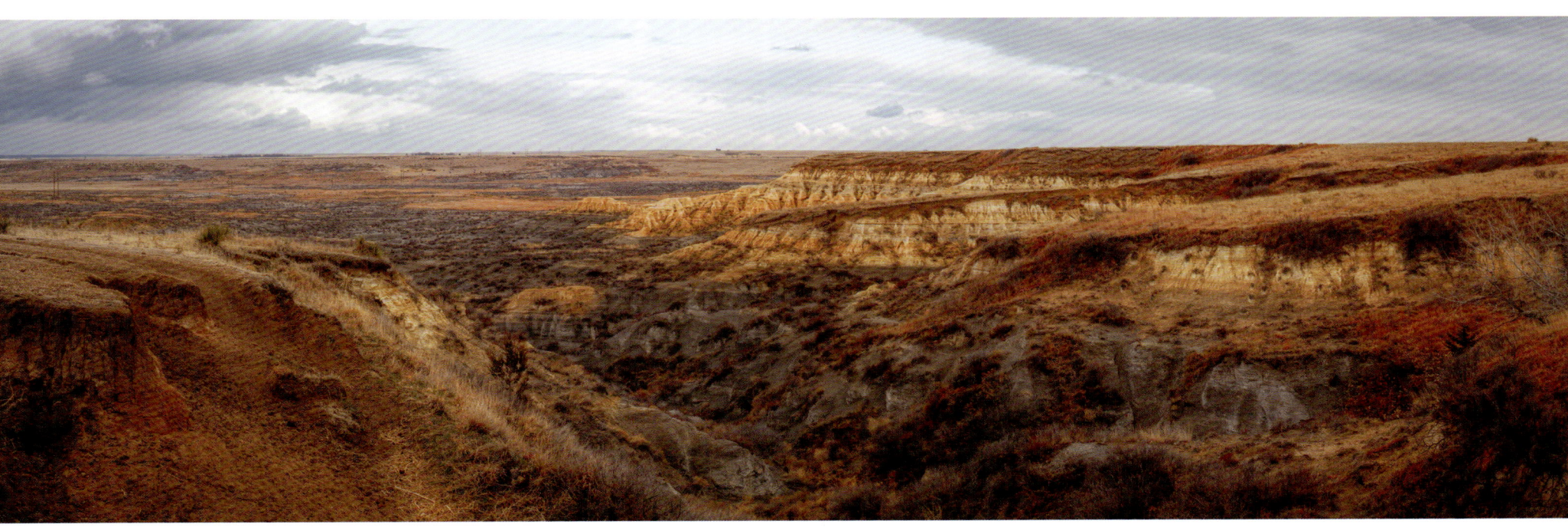

Rocks of the Smoky Hills: Return of the Sea

Geology of the Smoky Hills

The rocks of the Smoky Hills Province include sandstones, limestones, chalk, and shales that were deposited during the Cretaceous Period, which lasted from 145 to 66 million years ago. The earliest rocks of the Cretaceous are sandstones that were deposited on top of the late Permian rocks that are exposed to the east. Those late Permian rocks date to about 260 million years ago, which means that 115 million years of geological time are missing in Kansas. That time is represented by Triassic and Jurassic rocks that are present in states farther to the west. Those rocks include extensive eolian sandstones (windblown sand dunes) as well as river and floodplain deposits.

The reason for this gap in the rock record is that the midcontinent was exposed as land for most of that time, and any sediments that were deposited were removed later by erosion. Such erosional gaps in the rock record are called unconformities. At any given location, much of the geological history at that place is left unrecorded due to times of nondeposition and erosion.

The Smoky Hills themselves lie on the eastern side of the province and are composed largely of early Cretaceous sandstones deposited as beach sands and coastal river deposits. These sandstone hills gave their name to the entire province.

Lying above the sandstones, and exposed to the west, is an interval of alternating thin limestones and gray shales. One of these thin-bedded limestone units, called the Fence Post Limestone, was used by early settlers for buildings and fence posts because of the lack of available wood. The youngest Cretaceous rocks, forming the westernmost exposures, are shales and thick chalk beds. The resistant fossil-rich chalk beds have been eroded into bluffs along river channels and into badlands with mesas, flat-topped pillars, pinnacles, and irregular spires.

A Record of Rising Seas

The early Cretaceous sandstones record the initial flooding of the midcontinent by the ocean after a long period of exposure as land. The seas gradually spread across the midcontinent, leaving behind a record of the shoreline environments. Sandstones were deposited by coastal rivers and as shoreline sands along the margins of the rising sea. Associated with these sandstones were mudstones and clays accumulated in floodplains and swamps. Thin coals and clay-rich fossil soils indicate a warm, wet climate. These coastal deposits were followed by shallow marine limestones and shales.

The view at this location across some of the unique formations in Gove County really had an expansive feel for me. The layered sections of rocks on the right and the draw running through the photo combined to create visual interest. The distance to the horizon here also contributed to the feeling of space.

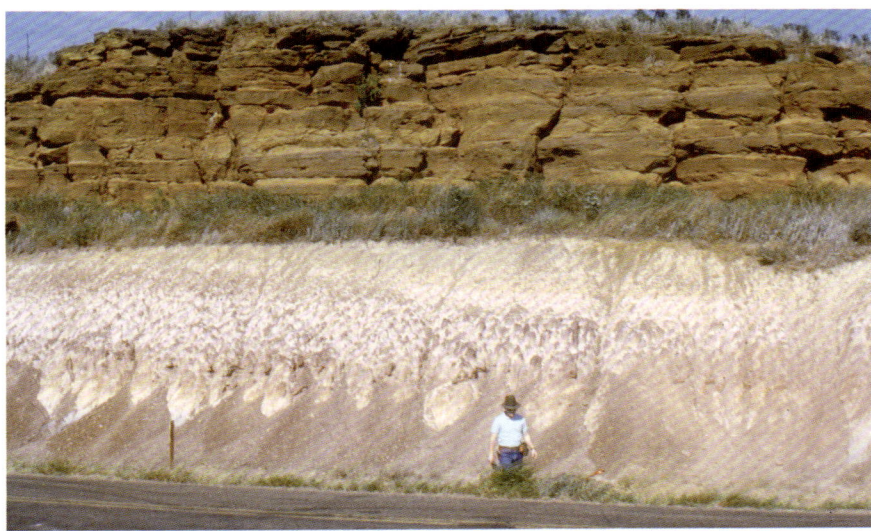

Above and right: This roadcut exposes the early Cretaceous Dakota Sandstone and the underlying Terra Cotta Clay. These rock units were deposited in nearshore settings as the seas began to rise and flood the midcontinent at the beginning of the Cretaceous. The Dakota Sandstone includes coastal river deposits and beach sands. Farther to the west in Colorado, the Dakota preserves dinosaur trackways. The red-and-yellow mottled Terra Cotta Clay is a fossil soil (paleosol) that was produced by intense weathering in a warm and wet climate. Such soils are mottled due to abundant iron oxide and have a high concentration of kaolinite clays, which are used in the making of pottery—thus the name "Terra Cotta Clay." (Dakota Sandstone and Terra Cotta Clay exposed along I-70 near the Brookville exit, Saline County. Photo by Keith Miller.)

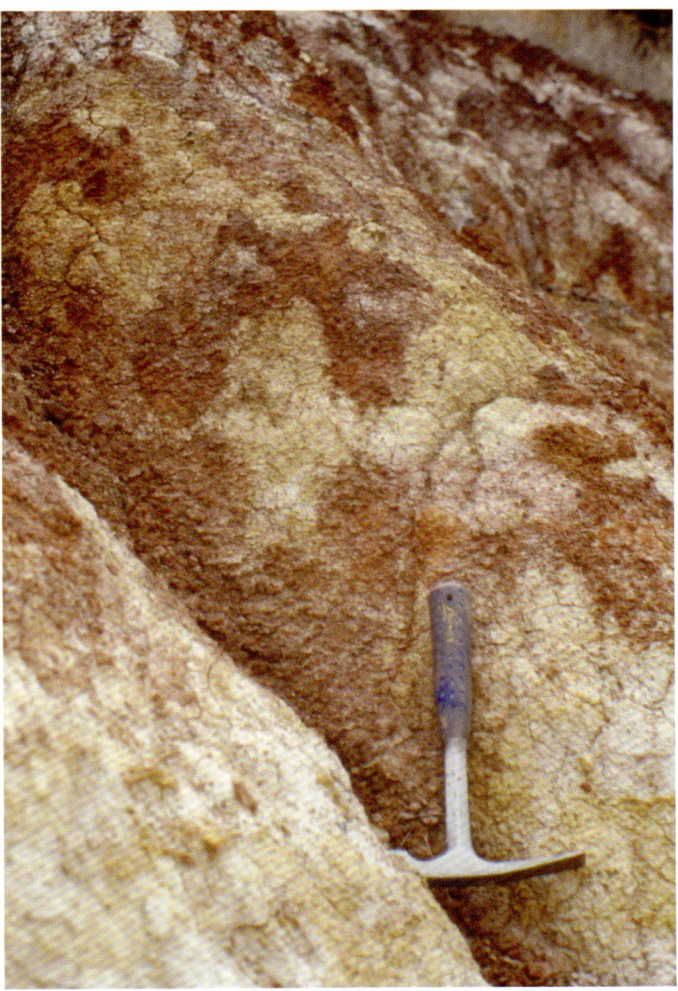

By the late Cretaceous, a shallow sea stretched across the midcontinent and northward across North America. At times, this Western Interior Seaway extended all the way to the Arctic and split the continent into two separate landmasses. When relative sea level fell, emergent land would temporarily rejoin the separated landmasses and their resident terrestrial species. In deeper water, very fine carbonate mud produced by single-celled planktonic algae called coccoliths covered the seafloor. Giant, flat, oyster-like clams up to six feet in diameter, called inoceramids, rested on this soupy mud, supported by their high surface area. These carbonate muds later hardened to form chalk.

The Cretaceous of Kansas provides an important record of marine life during this later period of the "Age of Dinosaurs." The river and lake deposits of the Triassic and Jurassic west of Kansas preserve abundant fossils of dinosaurs and of the earliest mammals. Discovery of these rich bone beds in the late 1800s led to intense competition among fossil hunters and an explosion of knowledge about terrestrial life in the Age of Dinosaurs. By contrast, Kansas provided an important record of ocean life, brought to light through the fieldwork of Charles Sternberg and his sons, among others. The fossil-rich late Cretaceous chalks preserve a wide variety of vertebrate fossils, including mosasaurs, plesiosaurs, marine turtles, twenty-foot-long *Xiphactinus* fish, aquatic toothed birds, and pterosaurs. These and other fossils discovered by the Sternbergs are displayed at the Sternberg Museum of Natural History in Hays.

The Cretaceous is also extremely important in the evolution of terrestrial plants. The first flowering plants (angiosperms) appeared in the earliest Cretaceous and rapidly diversified in the late Cretaceous. This dramatic change in vegetation, with the advent of flowers and fruit, had a huge impact on terrestrial ecosystems.

The end of the Cretaceous is marked by an astonishing catastrophic event—the impact of an asteroid off the northern coast of the Yucatán Peninsula around sixty-six million years ago. That event caused the second-largest extinction in Earth history, resulting in the elimination of three-quarters of all the animal and plant species on the planet. Terrestrial plant species, nonavian dinosaurs, pterosaurs, plesiosaurs, mosasaurs, ammonoid cephalopods, inoceramids, and even many types of marine plankton disappeared. This event defines the end of the Cretaceous and the beginning of the Cenozoic Era. The immediate effects of the impact included an enormous blast wave, global fires, and pressure waves and tsunamis that traveled up the Cretaceous inland sea as far north as North Dakota. Dust, blasted into the upper atmosphere, created several years of darkness and a "nuclear winter." When the dust cleared, high carbon dioxide concentrations in the atmosphere from the vaporization of limestone by the asteroid impact resulted in a rapid rise in global temperatures—a runaway greenhouse. This was not a good time to be alive! But certain groups of animals did survive, including a variety of small mammals, birds, amphibians, turtles, snakes, lizards, crocodilians, and bony fish. The surviving mammals and birds rapidly evolved and diversified after the extinction event to form the foundation of our modern vertebrate communities.

Smoky Hills Landscapes

The area around Kanopolis Lake in Ellsworth County exposes one-hundred-million-year-old Cretaceous rocks of the Dakota Formation and the underlying Kiowa Formation. The lower part of the Kiowa consists primarily of gray to black shales with thin limestone beds. The shales contain wood fragments and logs, as well as fossil marine mollusks (bivalves and gastropods), and were likely deposited in relatively quiet shallow marine environments. The upper Kiowa, which is exposed in the Kanopolis Lake area, is mostly sandstone that is locally colored reddish brown by iron oxide (hematite) staining. Unusual rose-shaped concretions called "barite roses," formed by barite crystals that grew into the surrounding sand, can be found in the area. These sandstones also display current-ripple marks and cross-bedding indicating faster-flowing water nearshore.

The prominent resistant sandstone beds that cap the hills in this area of the Smoky Hills belong to the Dakota Formation. These sandstones are commonly cross-bedded and likely record deposition by coastal streams and rivers. They are cemented with the iron oxide minerals of hematite and limonite, giving them a brown to yellowish-brown color. Interbedded with the sandstones are greenish-gray siltstones and red and reddish-brown mottled clays. The latter are paleosols formed under warm and wet climatic conditions. Furthermore, thin beds of low-grade coal (lignite) within the Dakota indicate swampy environments. Also present are fossil leaves—including some of the first common flowering plants. These include representatives of modern groups, such as viburnums, magnolias, laurels, and sycamores.

The Dakota Formation is exposed along the shoreline of Wilson Lake in

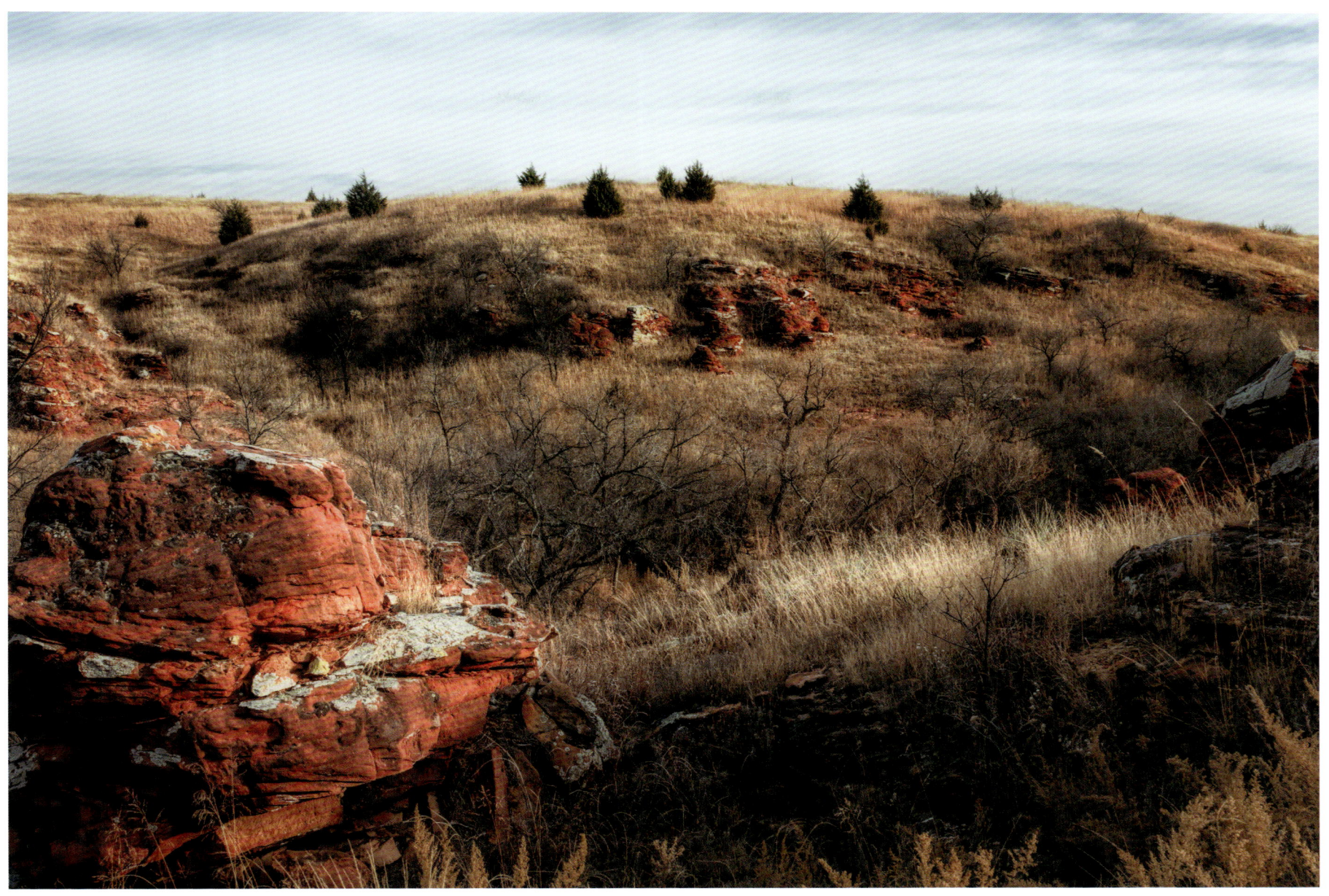

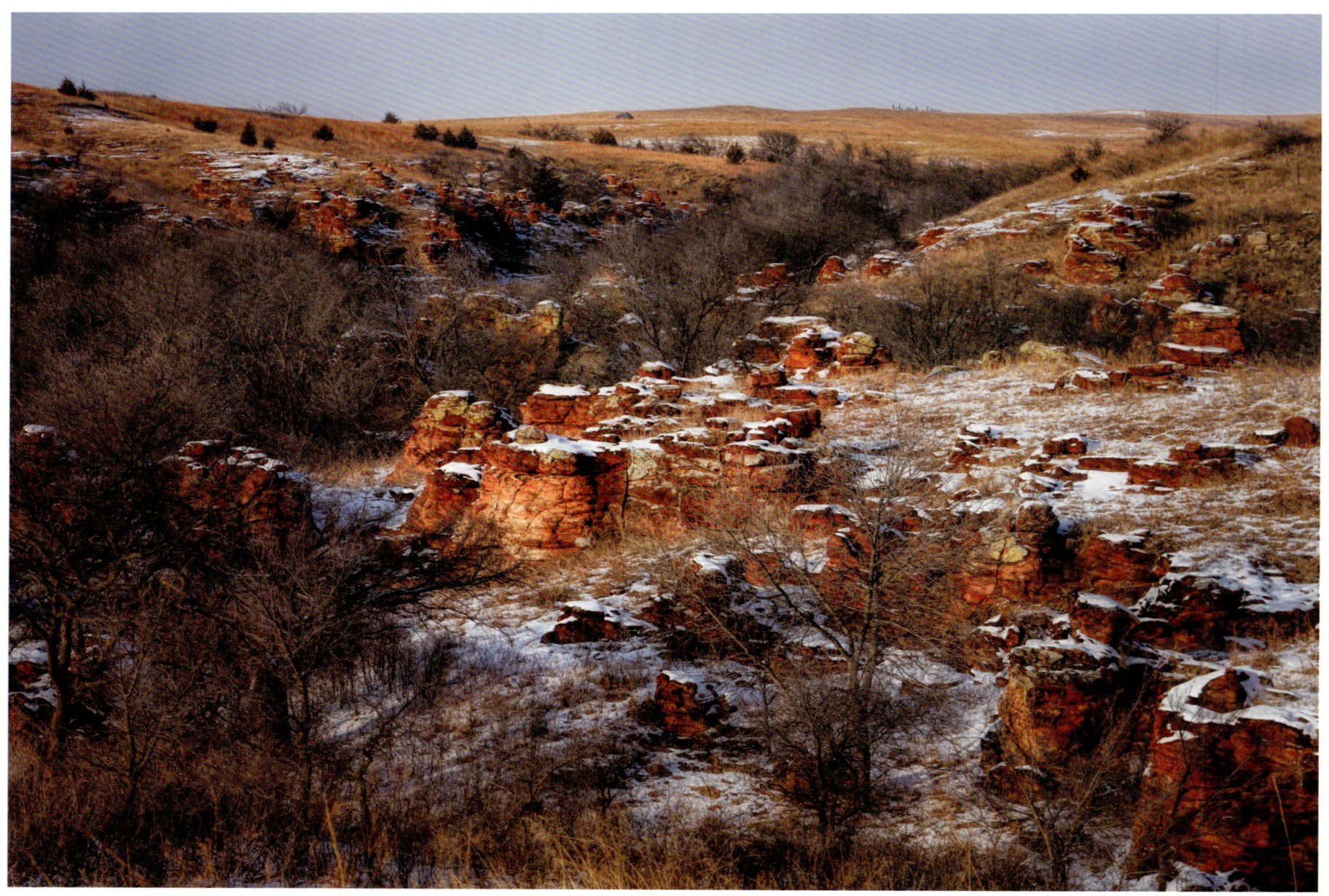

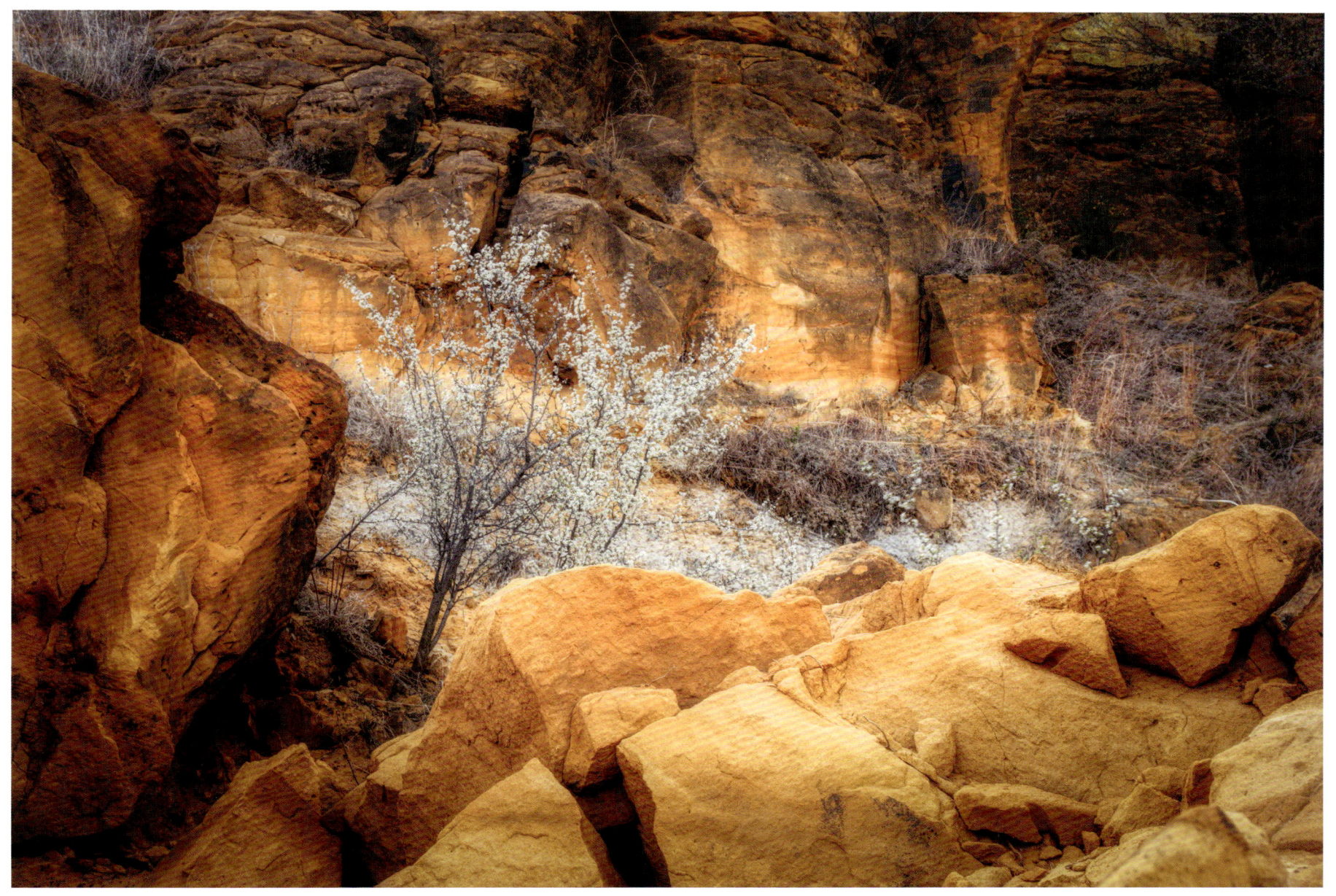

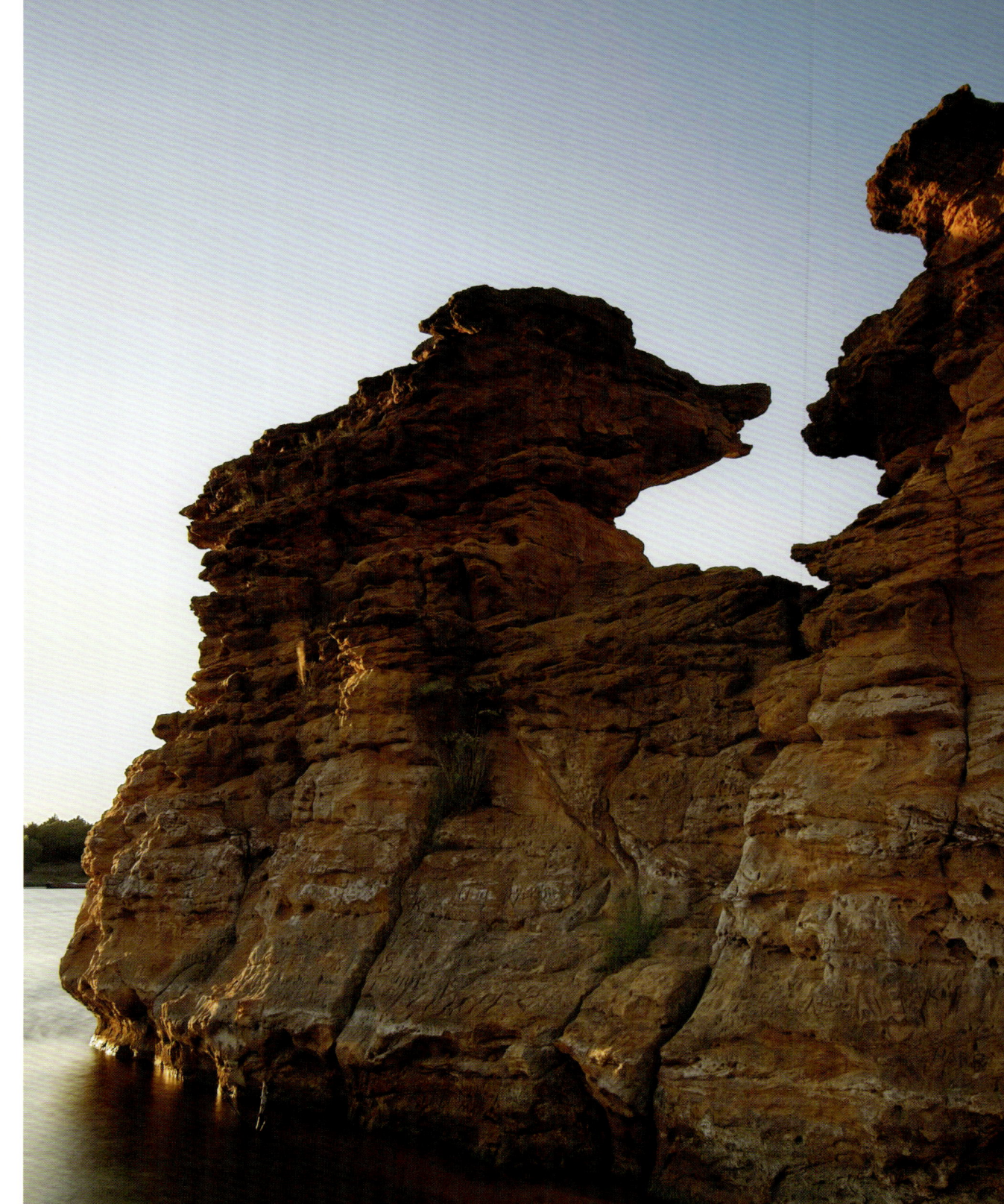

Page 69: The rock formations along the hiking trails at Kanopolis Lake offer a wide variety of textures and colors to photograph. On this late fall day, the red rocks matched well with the colors of the prairie.

Page 70: This photo is from a winter day at Kanopolis Lake. The repeating shapes of the rocks that disappear into the distance provide a visual clue to depth in this photo. The contrasting colors of the rocks and the light snow also added visual interest.

Page 71: The blossoms on the small bush nestled among the rocks caught my eye on this spring day at Kanopolis Lake. Areas along the trails sheltered so many bushes in bloom that you could smell them before you could see them. You could also hear the buzzing of the bees and other pollinators visiting the blossoms.

Right: Among the most popular destinations along Wilson Lake in north-central Kansas are the towering sandstone formations that rise out of the clear lake. These formations are accessible by boat or a two-and-a-half-mile trail and are visited by many people each day during good-weather months. I was lucky to visit on a weekday and a bit off-season but still had to share its beauty with others during my visit. (Photo by Mark Clarke.)

Russell County on the Saline River. At this locality, shark teeth and ammonite cephalopods (relatives of modern octopus and squid) can be found in some layers, indicating deposition in a shallow sea. Ammonite cephalopods were especially abundant during the Mesozoic and went extinct during the end-Cretaceous extinction event. Fossil plants are also found within concretionary nodules in the sandstone. The nodules were formed by the precipitation of minerals from water flowing through the sandstone, often around organic matter such as decaying plants. Thus, the Dakota here represents a transition between nearshore terrestrial settings and shallow ocean environments along the rising and expanding inland sea.

Large spherical concretions occurring in parts of the Dakota Formation were formed by the precipitation of calcite cement within a quartz sandstone. Initially forming around a small nucleus, the concretions grew outward concentrically within the sandstone, eventually reaching their giant size. The surrounding sandstone is only weakly cemented by iron oxide and is more easily eroded than the more resistant concretions.

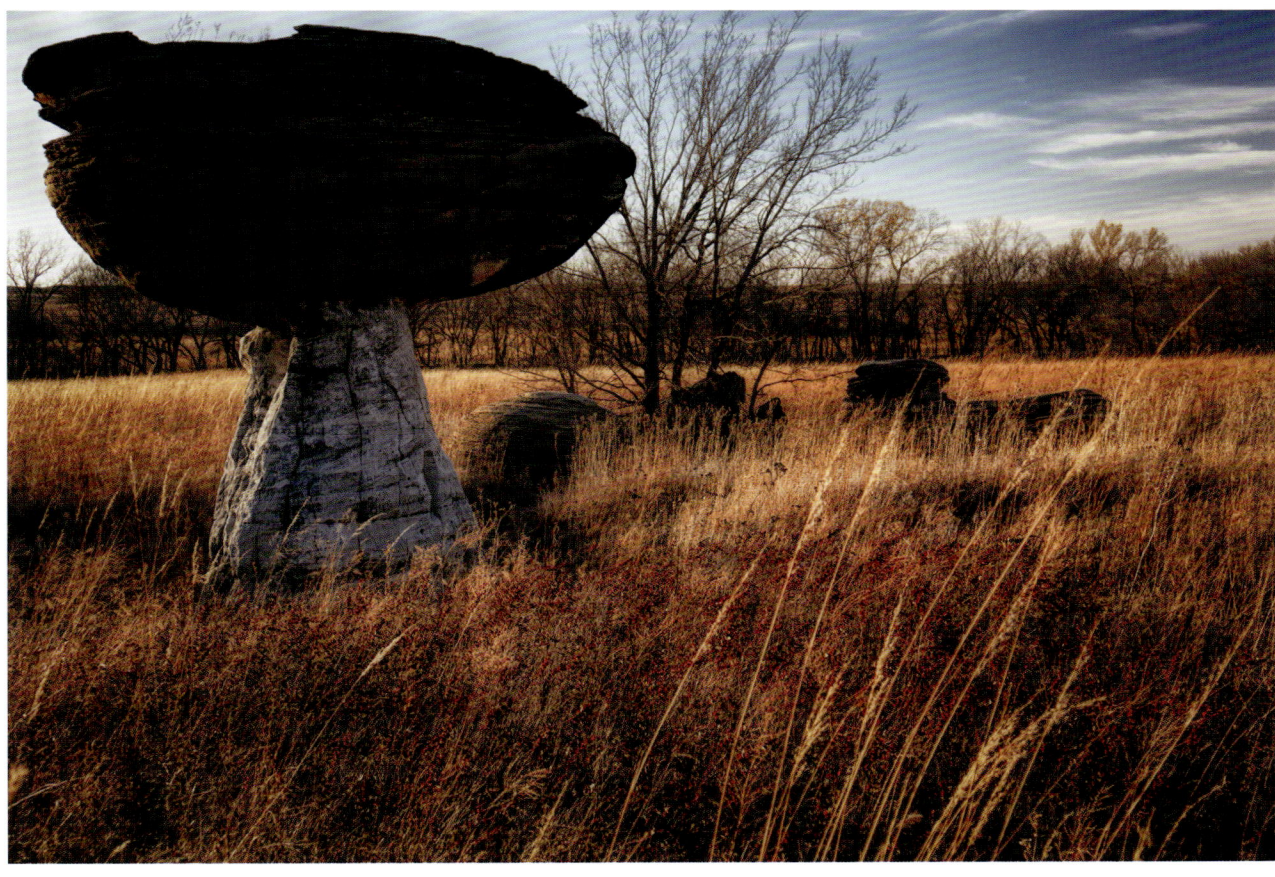

The balanced rocks at Mushroom Rock State Park in Ellsworth County seem impossibly delicate. It was difficult for me to wander around the rocks and not think about the processes that led to their creation.

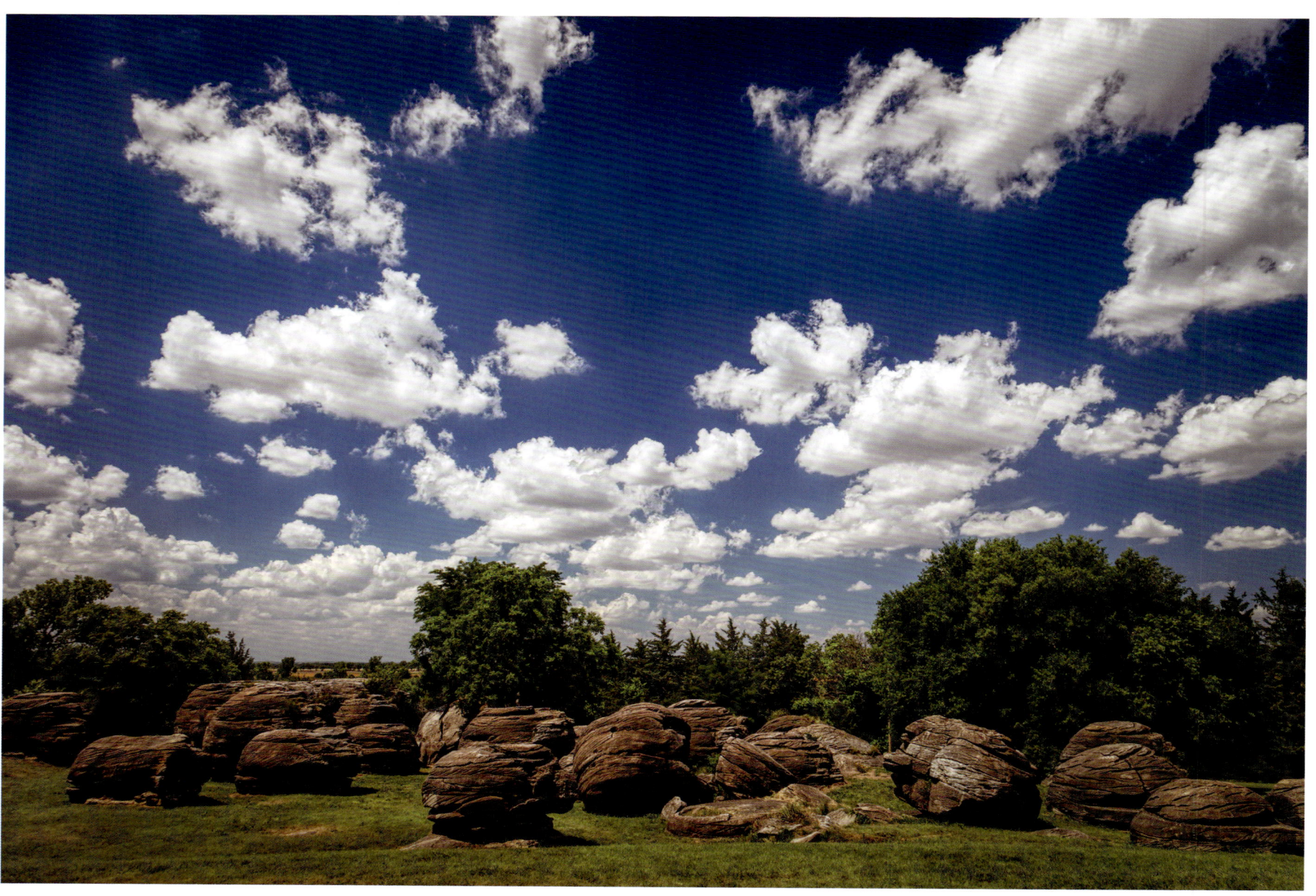

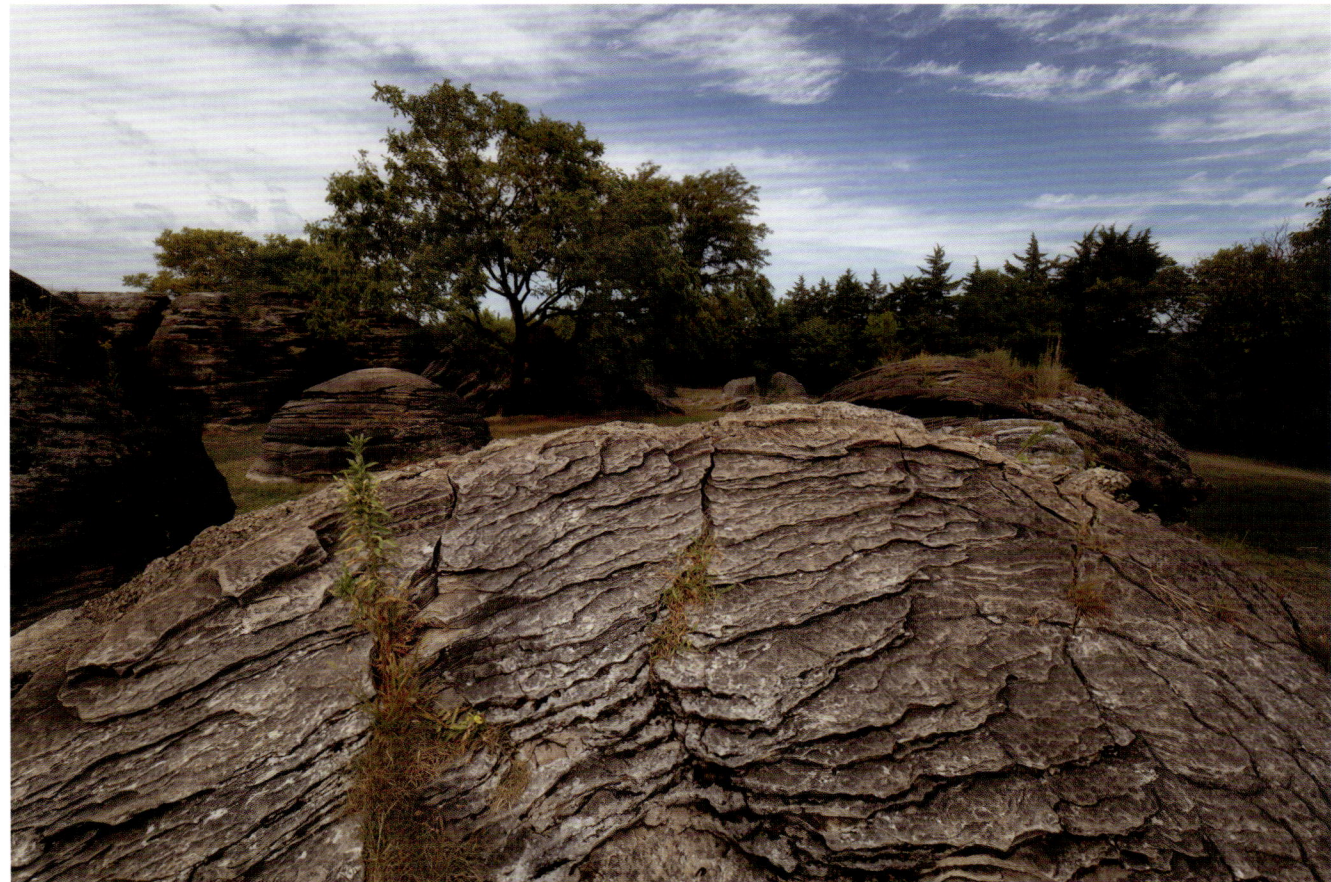

The concretions seen at Rock City near Minneapolis range from ten to twenty feet in diameter. The sandstone around the concretions has been completely eroded away, leaving the concretions resting on the ground like giant pool balls. Looking closely, you can see the cross-bedding and current ripples characteristic of river-deposited sediment. At Mushroom Rock State Park, some of the concretions sit on top of pedestals of the softer uncemented sandstone that has not yet been entirely eroded away.

The rocky cliffs of Cedar Bluff belong to the Fort Hays Limestone at the base of the Niobrara Chalk. The thick-bedded limestone is resistant to erosion and forms bluffs in counties where it is exposed. Composed of calcium carbonate microfossils produced by phytoplankton called coccoliths, the limestone is fine-grained and would have formed a soft, slurry-like mud on the shallow seafloor. Adapted to this soft, muddy bottom were the large, flat inoceramid clams that could "float" on this soupy surface. Fossil burrows indicate

Cedar Bluff Reservoir in Trego County was constructed in the early 1960s, mainly as a result of the Dust Bowl period in the 1930s. It is one of the most dramatic of the state parks in Kansas, with a 150-foot limestone bluff providing a scenic view of the entire area. (Photo by Mark Clarke.)

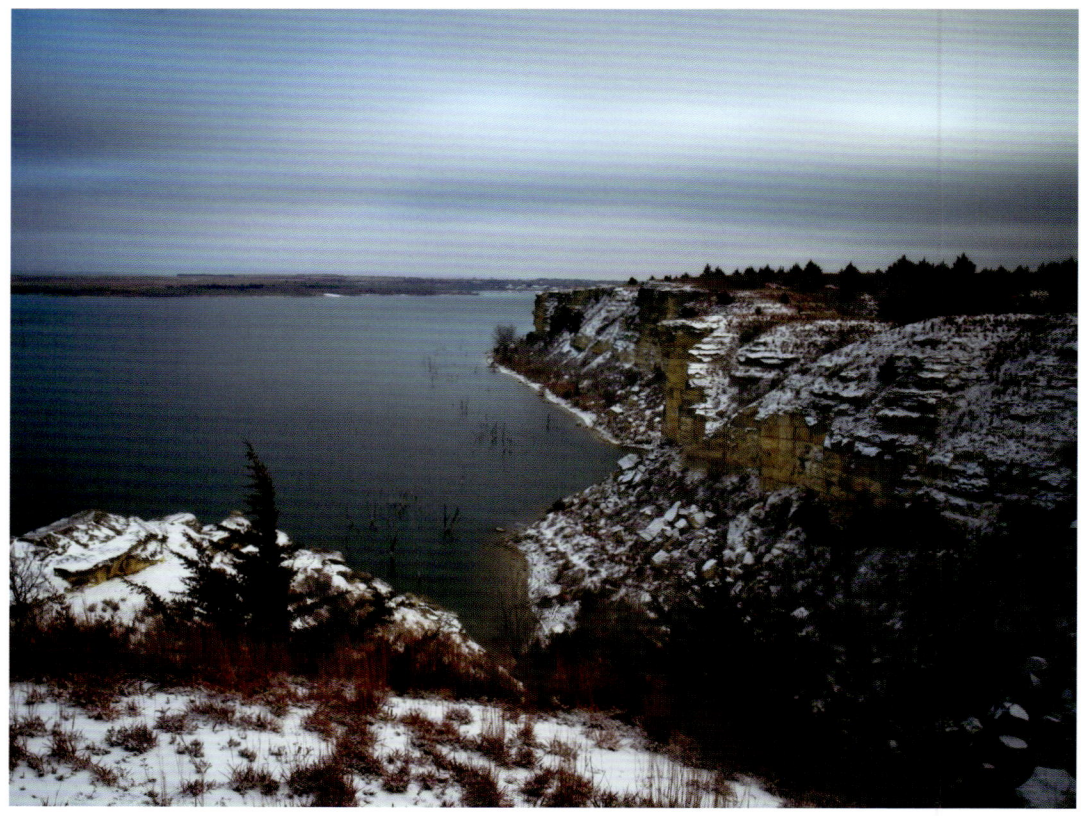

the presence of a variety of soft-bodied invertebrates living in the carbonate mud. Vertebrate fossils are not common but include sharks, large bony fish, and long-necked plesiosaurs that inhabited the warm waters above.

In Gove, Logan, and Trego Counties in western Kansas, remnants of the Smoky Hill Chalk (the uppermost part of the Niobrara Chalk) occur as flat-topped towers, pillars, and irregular spires scattered over the flat plains. This chalk once covered the area, but erosion has removed all but these distinctive features of the landscape. In the dry climate of western Kansas, water and wind erosion tend to cause vertical cliffs that retreat backward over time, producing flat-topped mesas and eventually remnant tower and castle-like features. The Smoky Hill Chalk includes both chalky limestone and chalky shale with distinctive thin horizontal layers. The flat-topped pillars at Monument Rocks and Castle Rock have more resistant cemented layers at the top that protect the softer layers below. Badlands are also present in Gove County where gullies and washes cut into the chalk layers below the surface. The badlands of Little Jerusalem in Logan County display the chalks in one-hundred-foot-high cliffs and spires that look like a city from a distance.

The thin, horizontal chalk layers were formed by the slow accumulation of calcareous coccolith skeletons on the seafloor at the very slow rate of about 0.0014 inches per year. The Smoky Hill Chalk is especially well-known for the spectacular fossil discoveries of fish, marine reptiles, and aquatic birds that inhabited the inland sea.

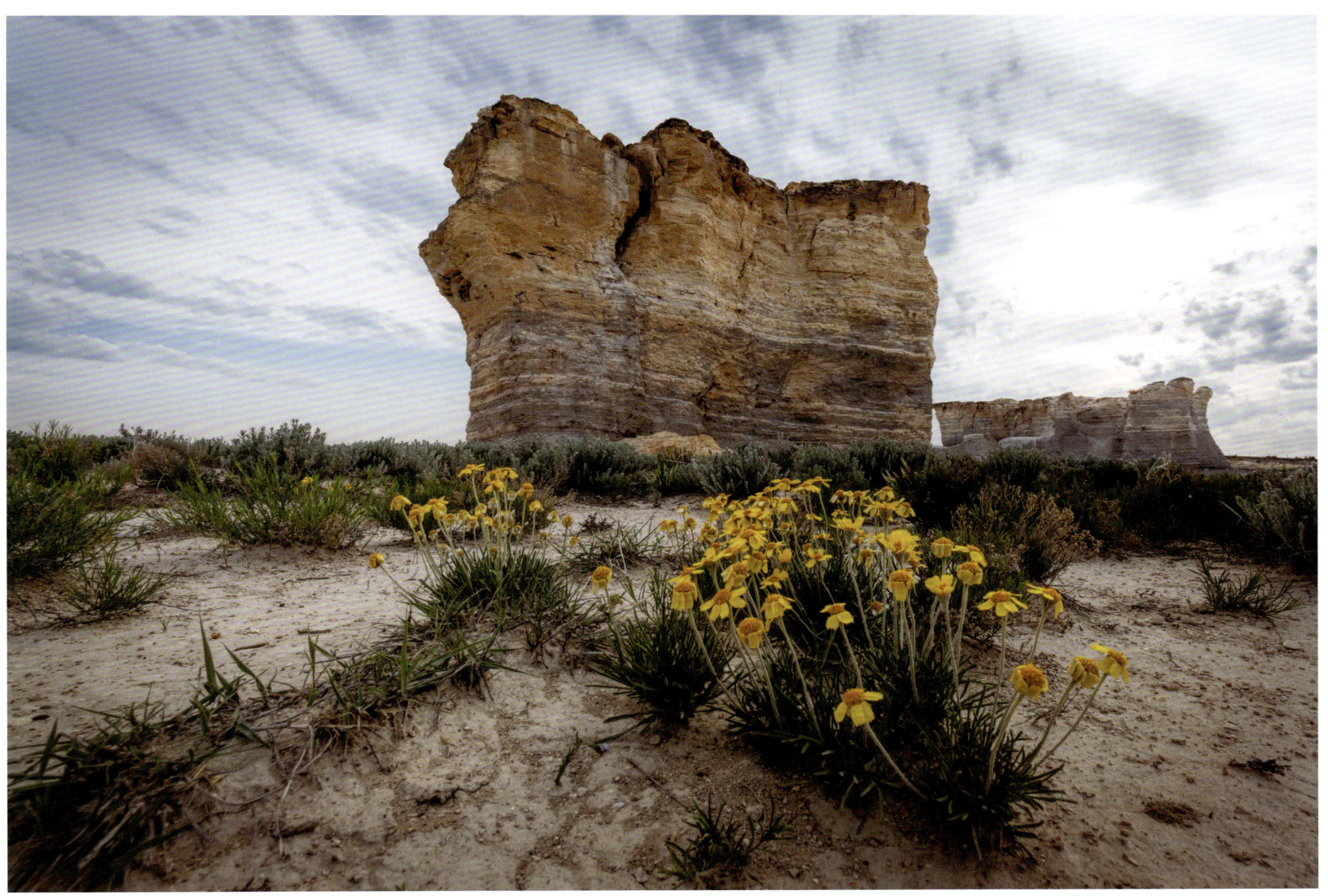

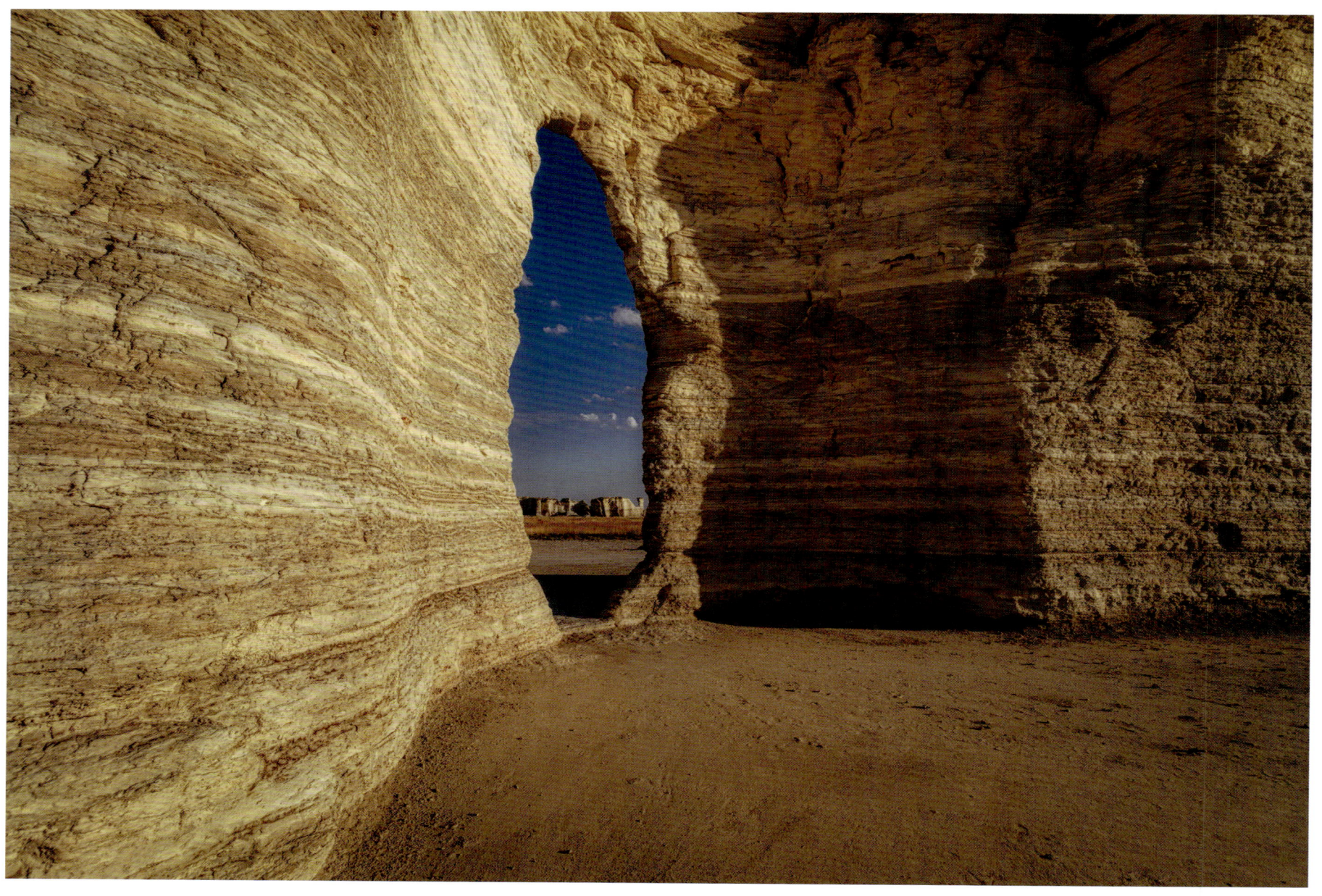

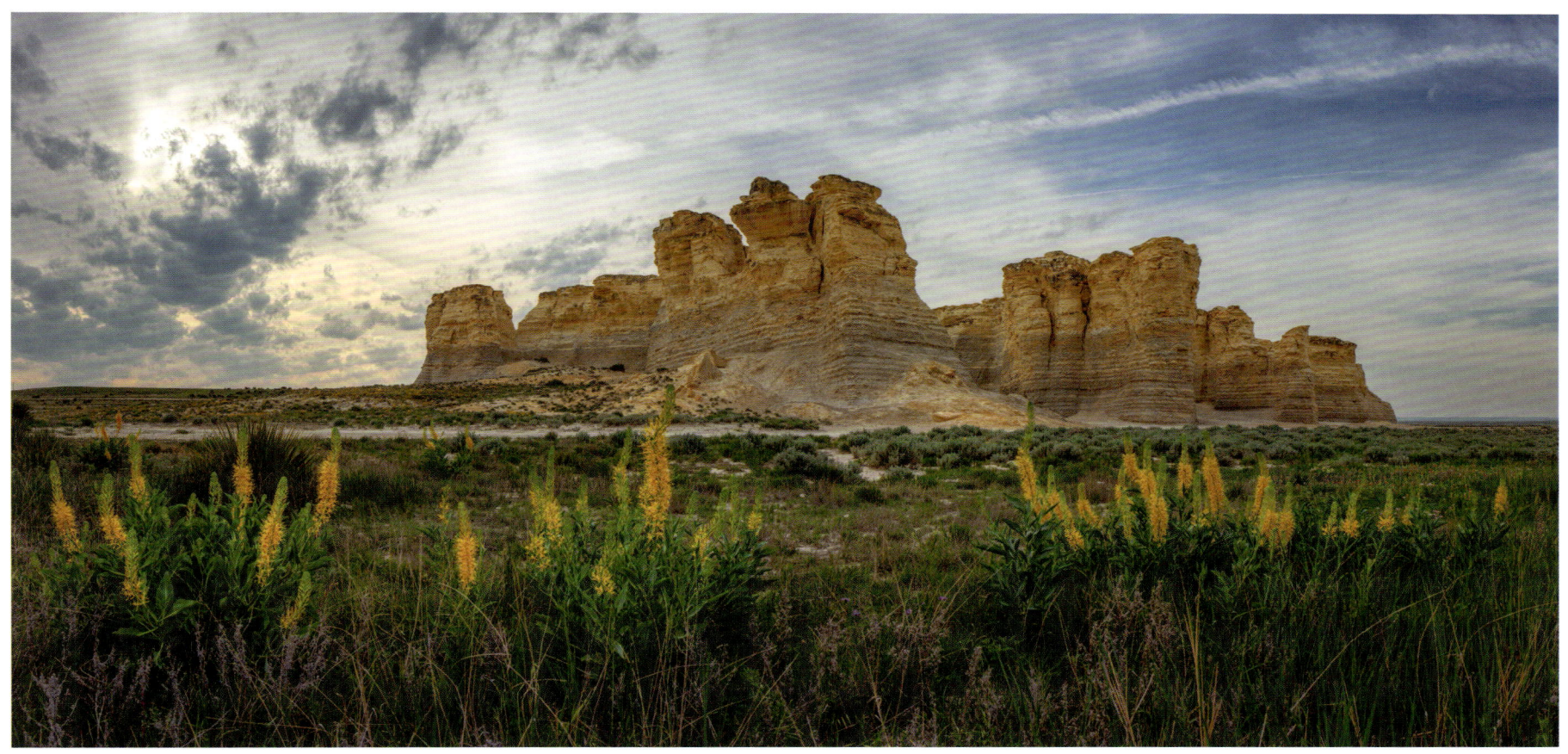

Page 77: Monument Rocks south of Quinter in Gove County is a well-known area and often photographed. However, there are a number of other interesting rock formations nearby, such as the one shown in this photograph. I liked the pairing of the rock with the flowers on this late spring day.

Page 78: I usually focus on the wide-open views of Kansas landscapes and tend to do most of my photography with a wide-angle lens. That type of photo conveys the feeling of openness and space that I often feel. However, the fine lines in the rocks shown in this photo from Monument Rocks grabbed my attention and made for a more interesting intimate photo that showed the detail of the chalk.

Above: Some of the rock formations found at the Castle Rock area in Gove County remind me of ships sailing across the prairie. The moment I made this photo was perfect, with the sky and wildflowers setting off the rock formation.

The textures and shapes of the rock formations at the Castle Rock area in Gove County made a perfect backdrop for this clump of wildflowers in the foreground.

Page 81: The fact that photography captures a moment that someone experienced is a valuable part of a photograph. Sometimes I can get lost in the technical parts of photography and forget to take time to experience those moments. The moment that this photograph captures was enjoyable as the storm built up in the background and the afternoon light revealed the details of the rock formations across the badlands at the Castle Rock area.

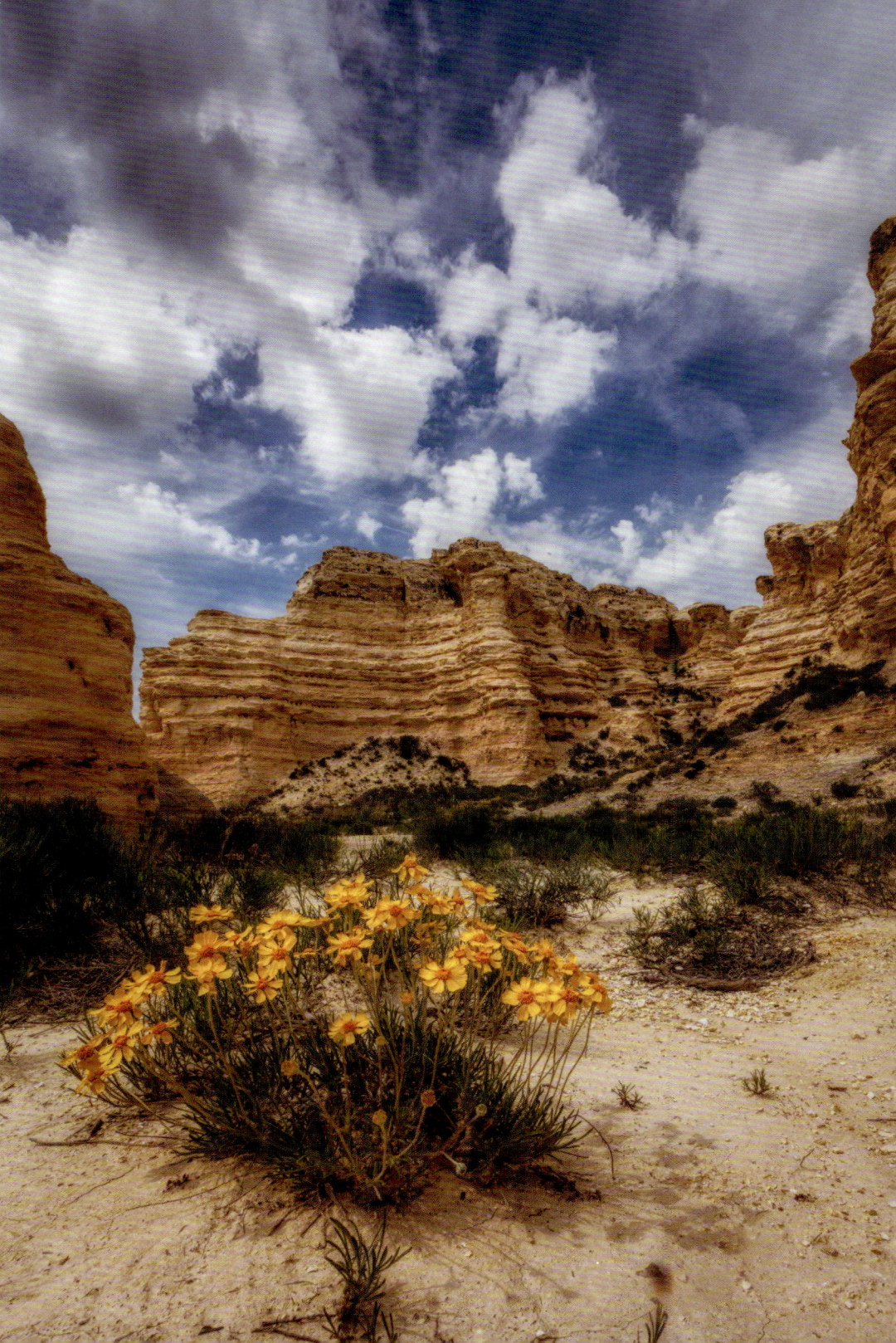

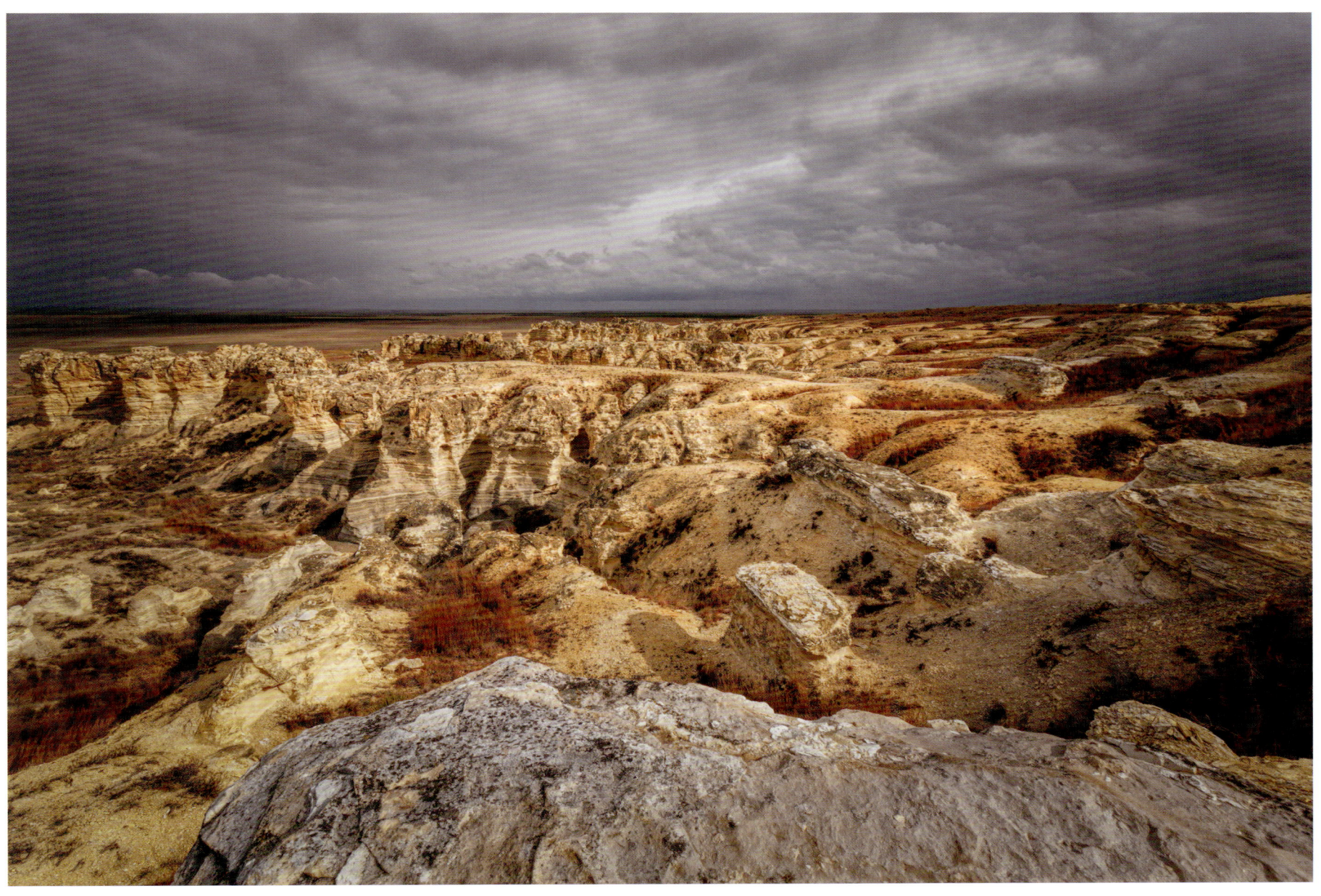

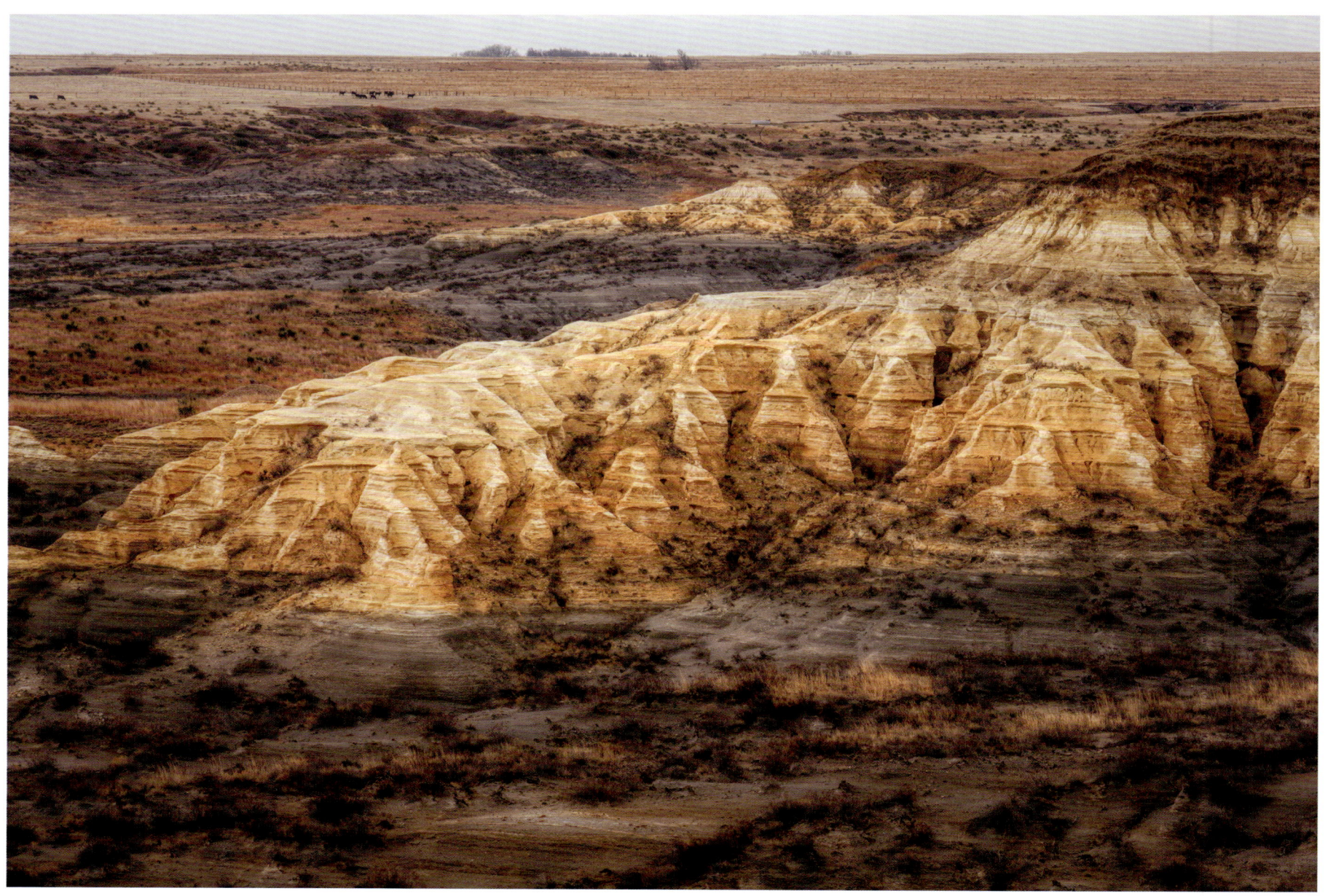

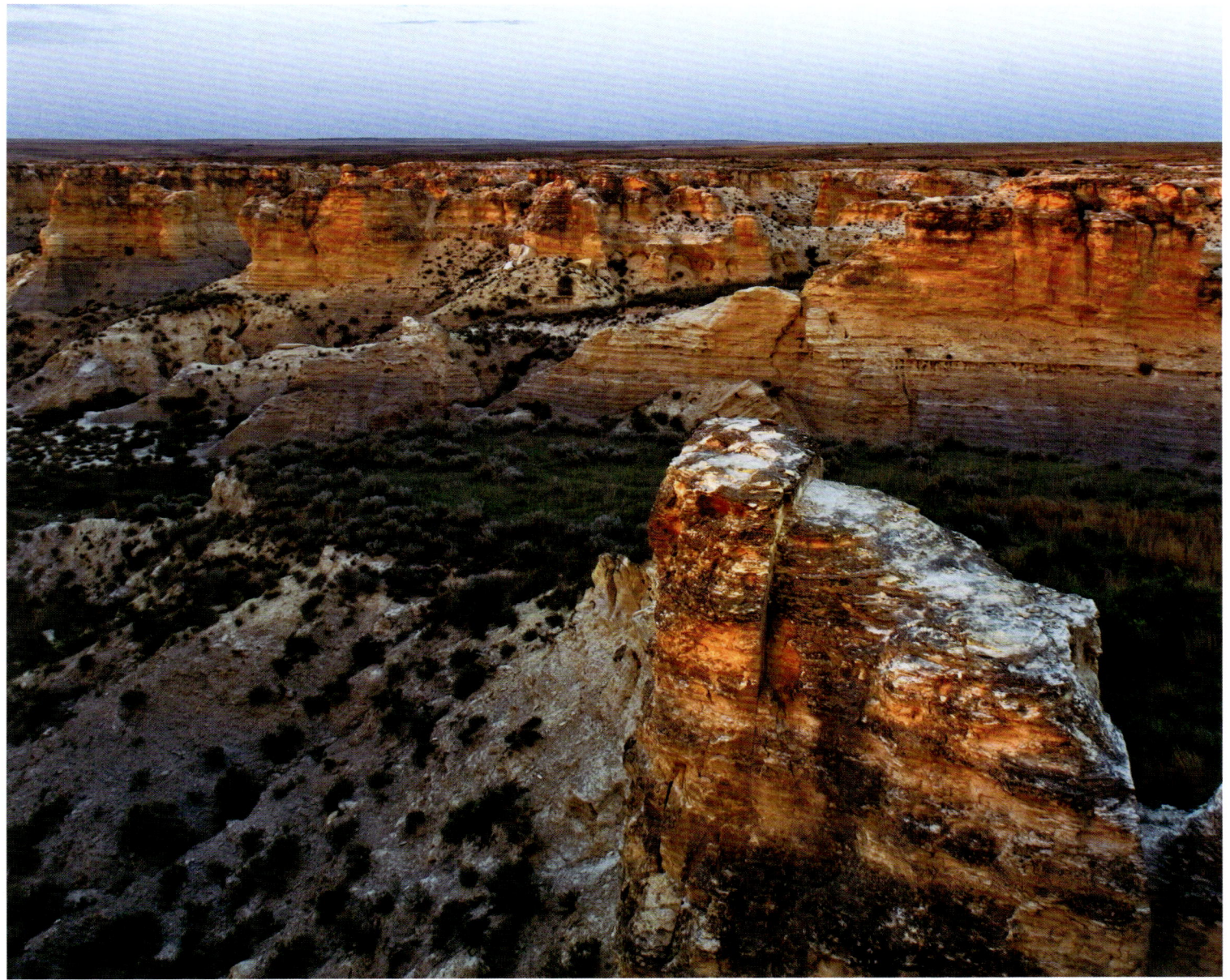

Page 82: These rock formations from Gove County looked a bit like they were slowly spreading out from the hillside. Seeing partially uncovered rock formations made me wonder what else is out there that we can't see.

Left: Little Jerusalem Badlands State Park in Logan County is the newest state park in Kansas and is owned by The Nature Conservancy and managed by the Kansas Department of Wildlife and Parks. These Niobrara Chalk formations, shown here in the orange glow of a sunset, are considered some of the most dramatic in western Kansas. (Photo by Mark Clarke.)

Section 6: High Plains

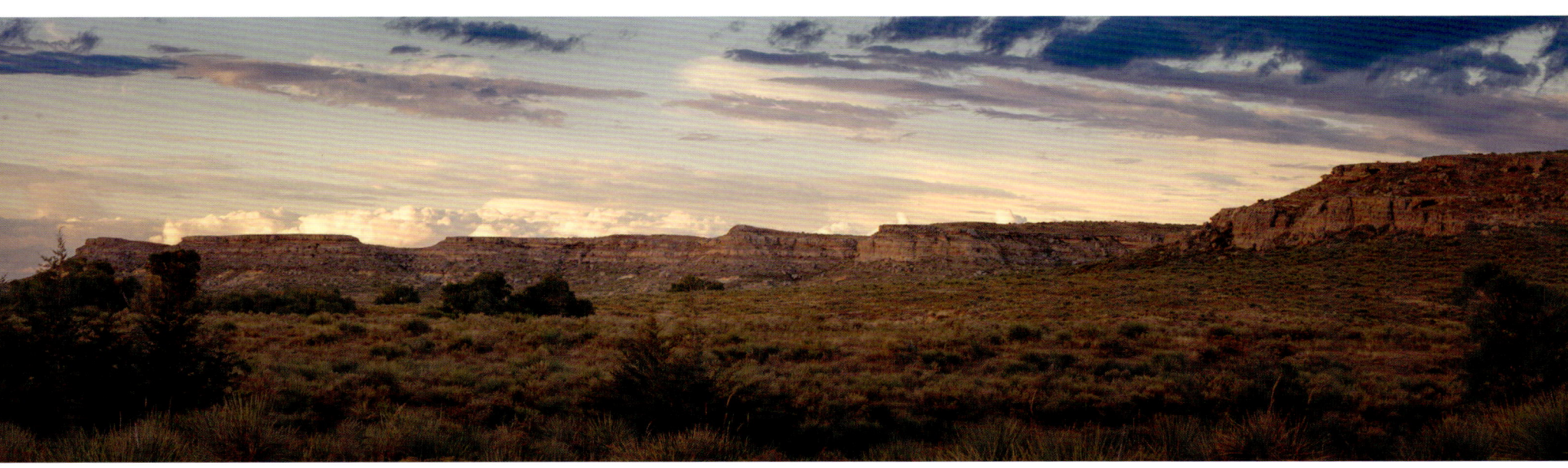

Rocks of the High Plains: Flatlands and Eroded Mountains

Geology of the High Plains

The High Plains is a region of flatland and gently rolling hills that covers nearly the entire western third of the state. Its name refers to its relatively high elevation. The land surface of Kansas rises from a low of 679 feet above sea level in the southeast corner of the state to its highest point near the Colorado border at 4,039 feet above sea level. The precipitation decreases from east to west, and the High Plains receive only about fifteen to twenty-five inches of annual precipitation. This semiarid climate is characterized by the shortgrass prairie ecosystem, with drought-tolerant grasses along with yucca and cacti. This is in contrast to the wetter tallgrass prairie to the east in the Flint Hills. The shortgrass prairie was occupied by the Kiowa, Comanche, and Arapahoe peoples, who hunted vast herds of bison and pronghorn.

The flatland of the High Plains is covered by the Ogallala Formation, which overlies the eroded and weathered top of the Cretaceous. The Ogallala Formation is of Miocene Age (about twenty-three million to five million years ago) and consists of sand, gravel, silt, and other rock debris eroded from the Rocky Mountains to the west. Much of this sediment is loose and unconsolidated. However, some localized layers are cemented by calcite precipitated from shallow lakes and from calcium-rich water moving through the sediment. These concrete-like layers are called caliche and form the resistant caprock of local

Previous page: I wanted to showcase the rock formations in the area north of Scott County Lake in this panoramic photo but at the same time capture a sense of place. The prairie included in the foreground helped to convey this better than a close-up alone. The timing was fortunate with the storm building in the distance hinting at the change of weather that would soon be moving across the landscape.

bluffs. There are also ash layers that were deposited from distant erupting volcanoes. Some of the silica from these ash beds dissolved and precipitated to form agate and chert.

The porous and permeable sediments of the Ogallala Formation, which are up to seven hundred feet thick, form a huge aquifer system in the midcontinent that stretches from South Dakota to Texas. The groundwater stored in these sediments provides about 30 percent of all irrigation water in the United States. Unfortunately, water has been withdrawn at rates far exceeding the rate of replenishment since the 1950s. This has become an issue of serious concern as the groundwater levels continue to decline. In southwest Kansas, groundwater levels have dropped more than thirty-five feet since 1996.

Covering the Ogallala Formation over broad areas is windblown silt, called loess. These fine sediments were carried by the wind from the margins of the Pleistocene continental glaciers to the north.

The End of the Inland Sea and the Rise of Mountains

Beginning in the latest Cretaceous, about eighty million years ago, the modern Rocky Mountains began to be uplifted. By the end of the Cretaceous, the inland seas had begun their final retreat, ending the last marine flooding of the continental interior. The now exposed floor of the Western Interior Seaway was elevated several thousand feet above sea level and subjected to weathering and erosion. Sediments eroding from the rising Rocky Mountains to the west were carried eastward by many slow-moving, meandering rivers that dumped sand and gravel on the eroded plains. By the Miocene, the Rocky Mountains were eroding down, and a thick wedge of coarse sediment was deposited over western Kansas as the Ogallala Formation.

The Miocene sediments of Kansas and the midcontinent contain a diverse vertebrate fossil record. These include gomphotheres (primitive elephants), camels, horses, chalicotheres (strange herbivores with horselike heads, long forelimbs, and clawed feet), rhinocerids, and giant piglike entelodonts as well as "bone-crushing" canid carnivores. These lived in a subtropical grassland savanna with scattered trees such as hackberries, elms, and walnuts. A warmer and drier climate in the midcontinent in the Miocene resulted in the expansion of grasslands characterized by warm-season grasses. This important ecological transition stimulated the rapid diversification of the grazing odd-toed and even-toed ungulates (hoofed herbivores).

High Plains Landscapes

Within the High Plains, the river deposits of the Ogallala and the loess deposits of the Pleistocene were laid down over the eroded surface of Permian and Cretaceous rocks. Although forming a flat landscape, the plains of the region are dissected by deep gullies and canyons that cut down into these hidden, much older rocks. In Clark County, these gullies expose the late Permian red sandstones and siltstones that comprise the Red Hills to the east as well as the early Cretaceous shales of the Kiowa Formation. In Lane County, gullies cut through the relatively thin layer of overlying loess and reveal the chalks of the Niobrara underneath.

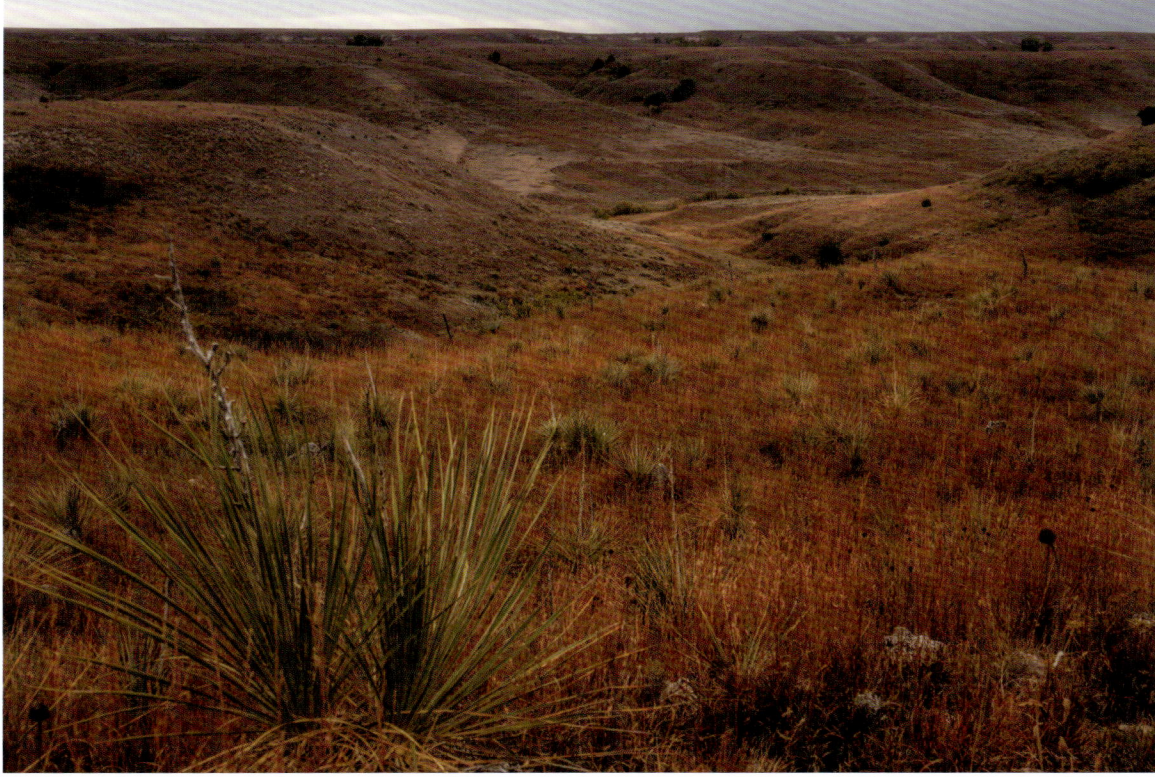

Draws like this one from Clark County Lake have always caught my eye. When the light is just right, shadows help bring out the shapes of the hills. The yuccas dotting the landscape provided a nice contrast to the prairie grasses in their fall colors.

Over the millennia, normally dry streams have cut shallow (ten to twelve feet deep) and wide channels into the otherwise mainly flat prairie of Lane County. After discovering these unique features, I knew I had to visit and attempt to capture this subtle landscape. (Photo by Mark Clarke.)

Opposite: Sunk below the rich farmland of Lane County and completely hidden from view, the shallow, dry stream-cut valleys reveal a wonderful array of colorful shale and limestone layers. These wide valleys are unknown to most people, even those who live close by. (Photo by Mark Clarke.)

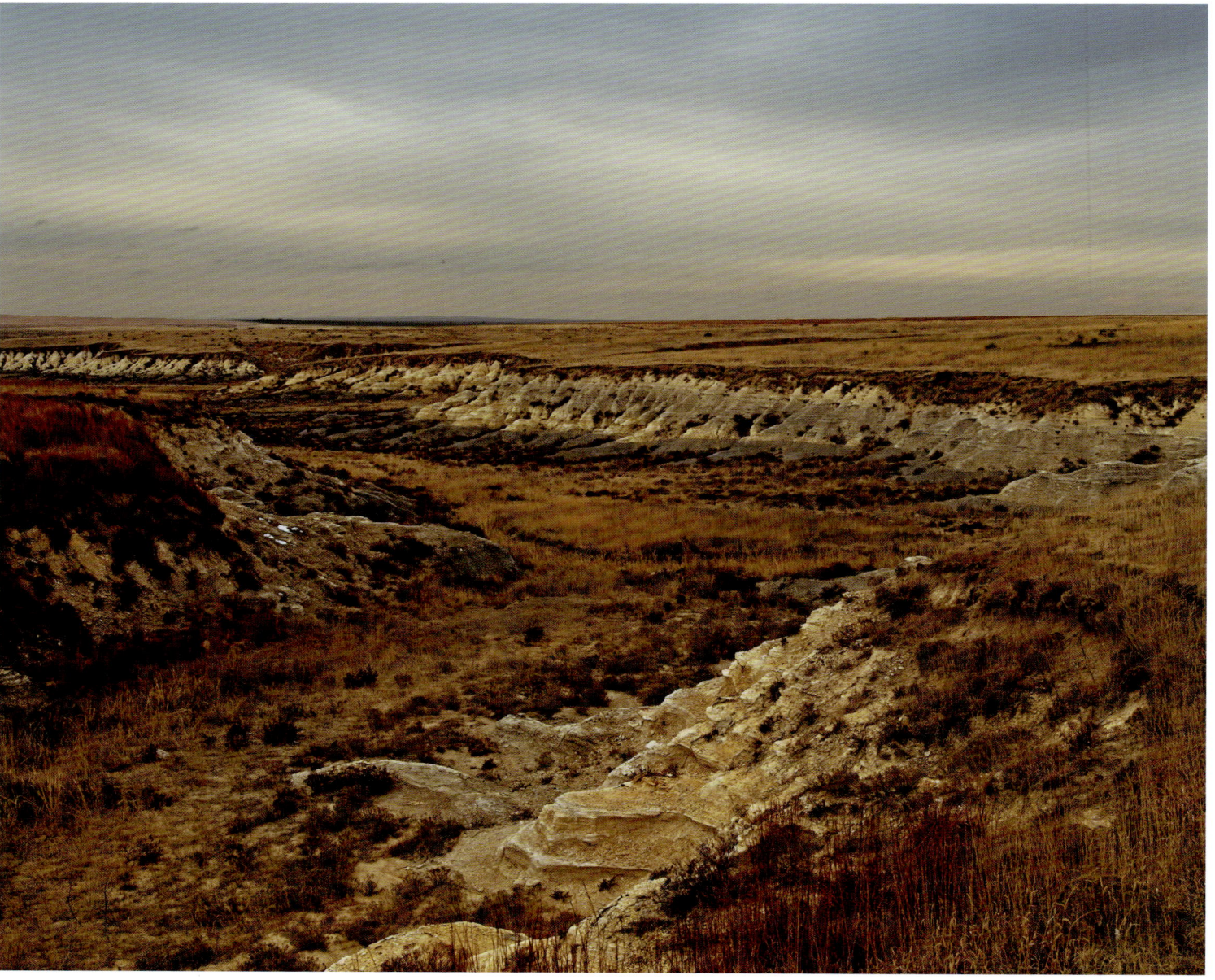

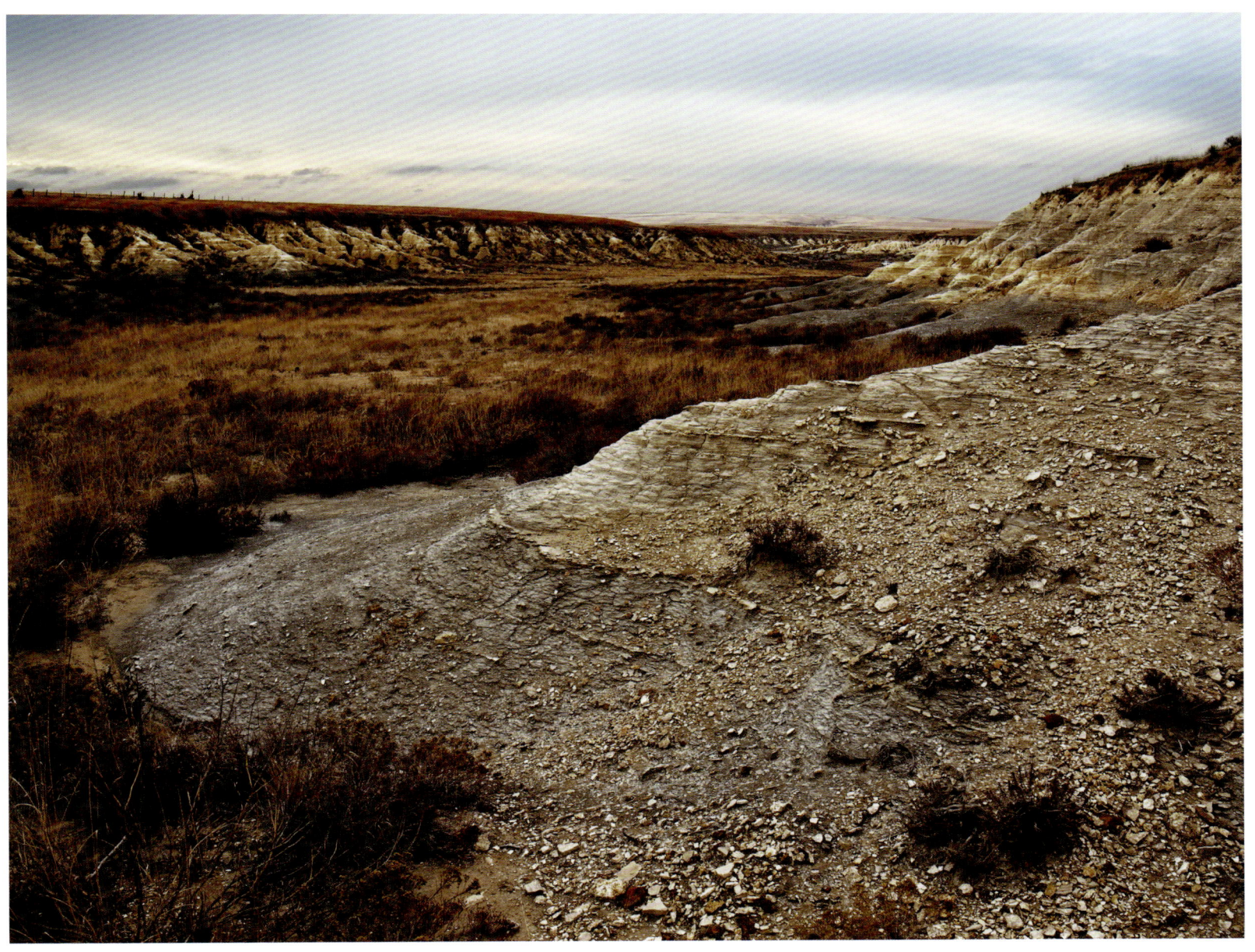

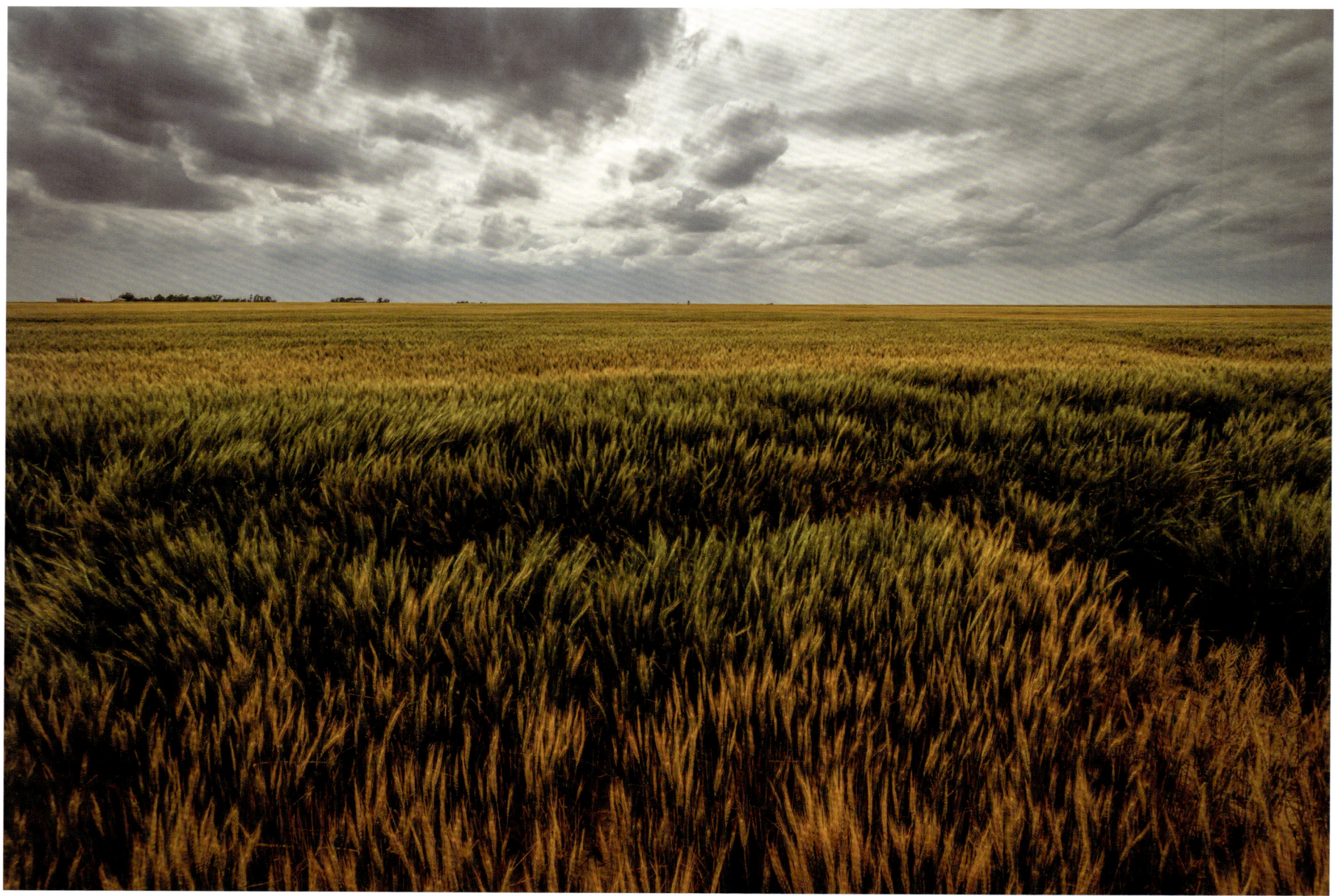

Opposite: A stormy day has the wheat swirling from side to side in this field in Lane County. I've always enjoyed watching the wheat dance in the wind when I'm standing at the edge of a Kansas wheat field. The movement and sound of the wheat swaying in the wind are mesmerizing.

Battle Canyon is named after the 1878 Battle of Punished Woman's Fork, which was an attack by US troops on a group of Northern Cheyenne under the leadership of Chief Dull Knife and Little Wolf, who were trying to return to their homeland in the north after escaping from a reservation at Fort Reno, Oklahoma. During the battle, the women, children, and elders were hidden in a cave at the top of the canyon. Escaping at night, the Cheyenne fled to Nebraska and some as far as Montana, but all were ultimately killed or captured.

The canyon is eroded into the calcite-cemented sands and gravels of the Ogallala Formation. These rubbly caliche layers, looking something like concrete, are resistant to weathering in the dry climate of western Kansas. Over time, however, water can dissolve the calcite and form caves, such as those used by the fleeing Cheyenne.

The river deposits of the Ogallala form a thickening wedge to the west toward their sediment sources in the rising Rocky Mountains. The photographs from Scott County show bluffs of the Ogallala that are capped by the

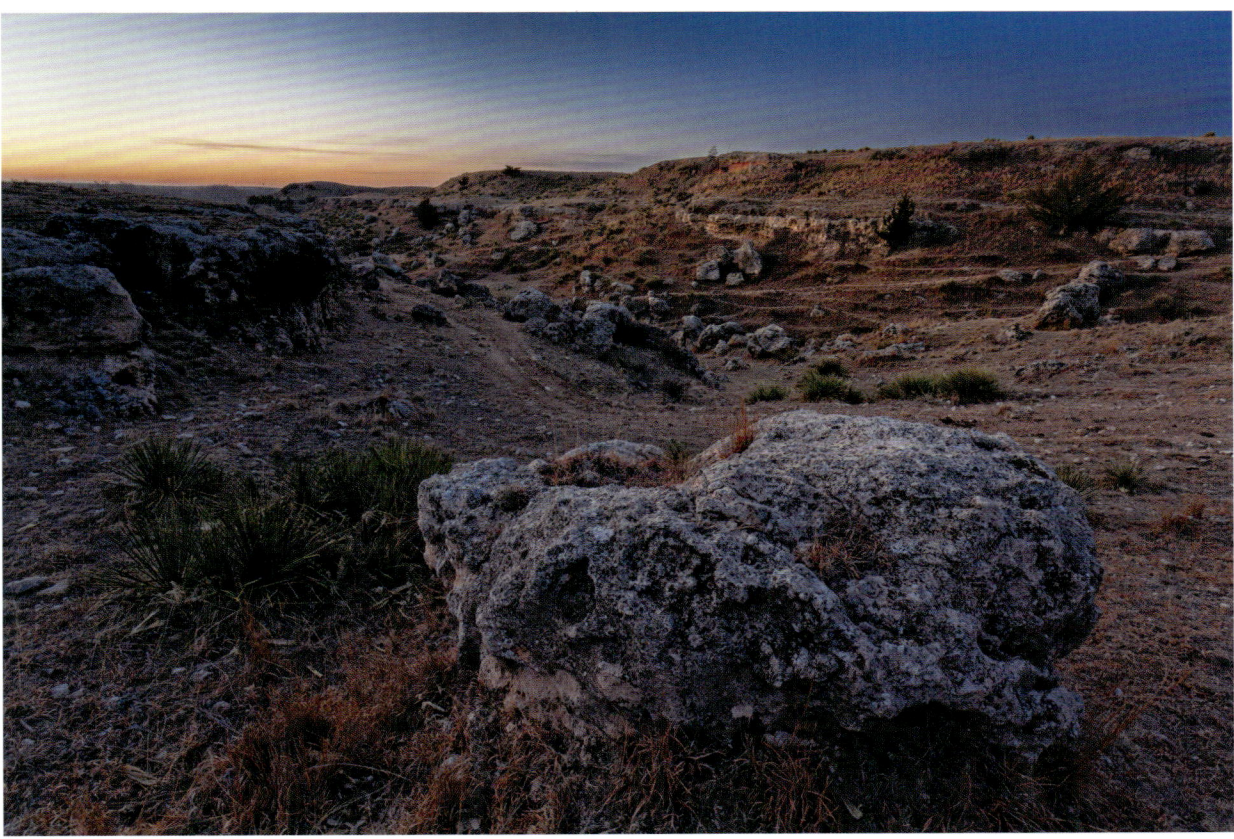

I took this photo just as the sun was rising at Battle Canyon near Scott County Lake. You can see the sun just starting to light up the rock formations in the background as it rises over the horizon just out of the left side of the frame. The lighter-colored rock in the foreground provided a nice contrast to the rest of the image as well as giving some texture to the photo. Many times when I'm composing a photograph, I select a location where you could sit and enjoy the view. The rock in the foreground would provide a good place to do just that.

This photograph from one of the edges of Battle Canyon shows the depth of the canyon and the lines of rock ridges along the sides. The afternoon sun provides more direct lighting, and the stronger light brings more contrast to the landscape, with darker shadows and brighter high-lights, than early-morning or late-evening light.

Opposite: There are many elements of a landscape to convey in a photograph—distance, shapes, colors, etc.—but texture is also an attribute to consider. I liked the textures provided by the different plants in this view of the plains near Scott County Lake in western Kansas. The rock formations on the horizon help to provide a sense of distance as well.

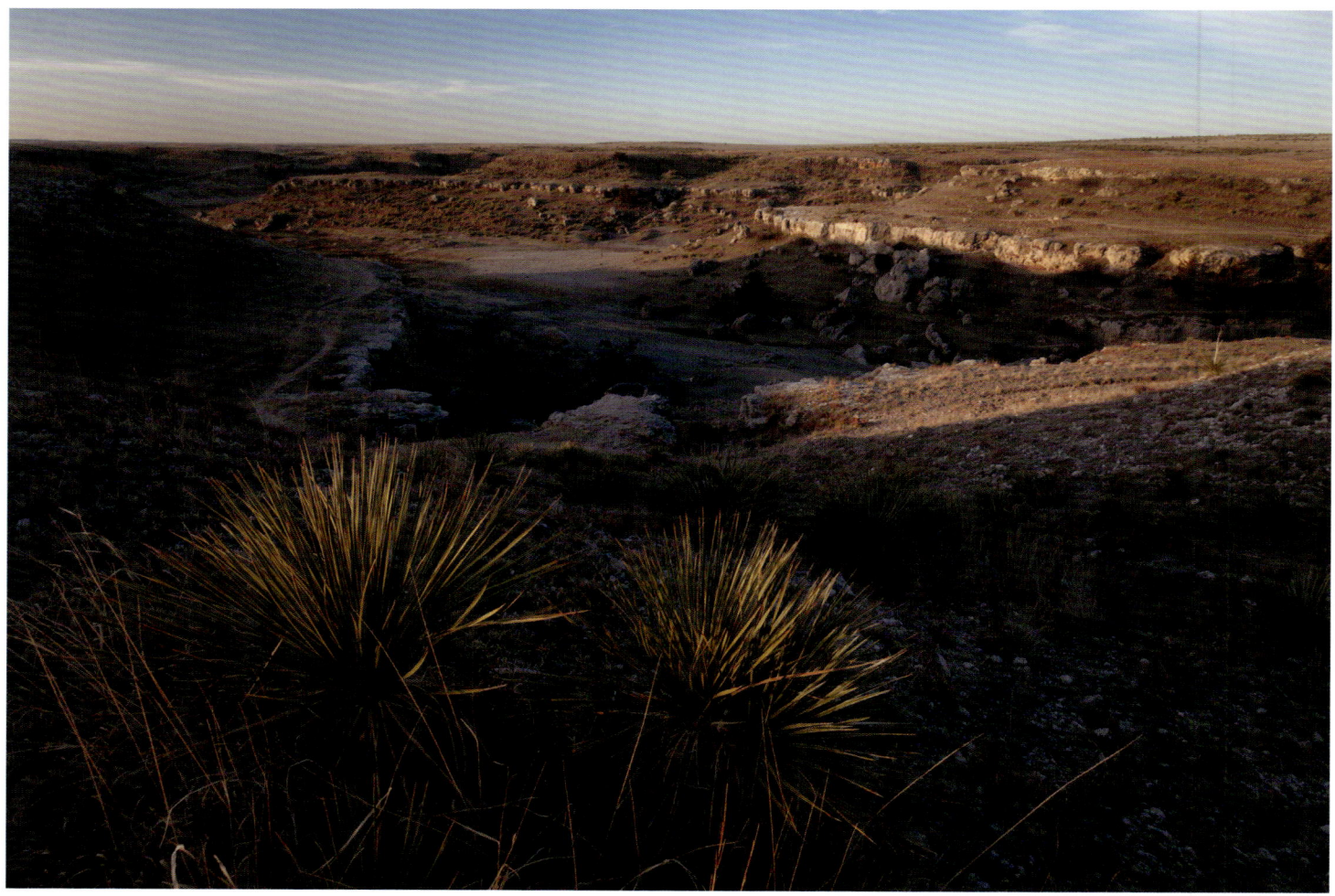

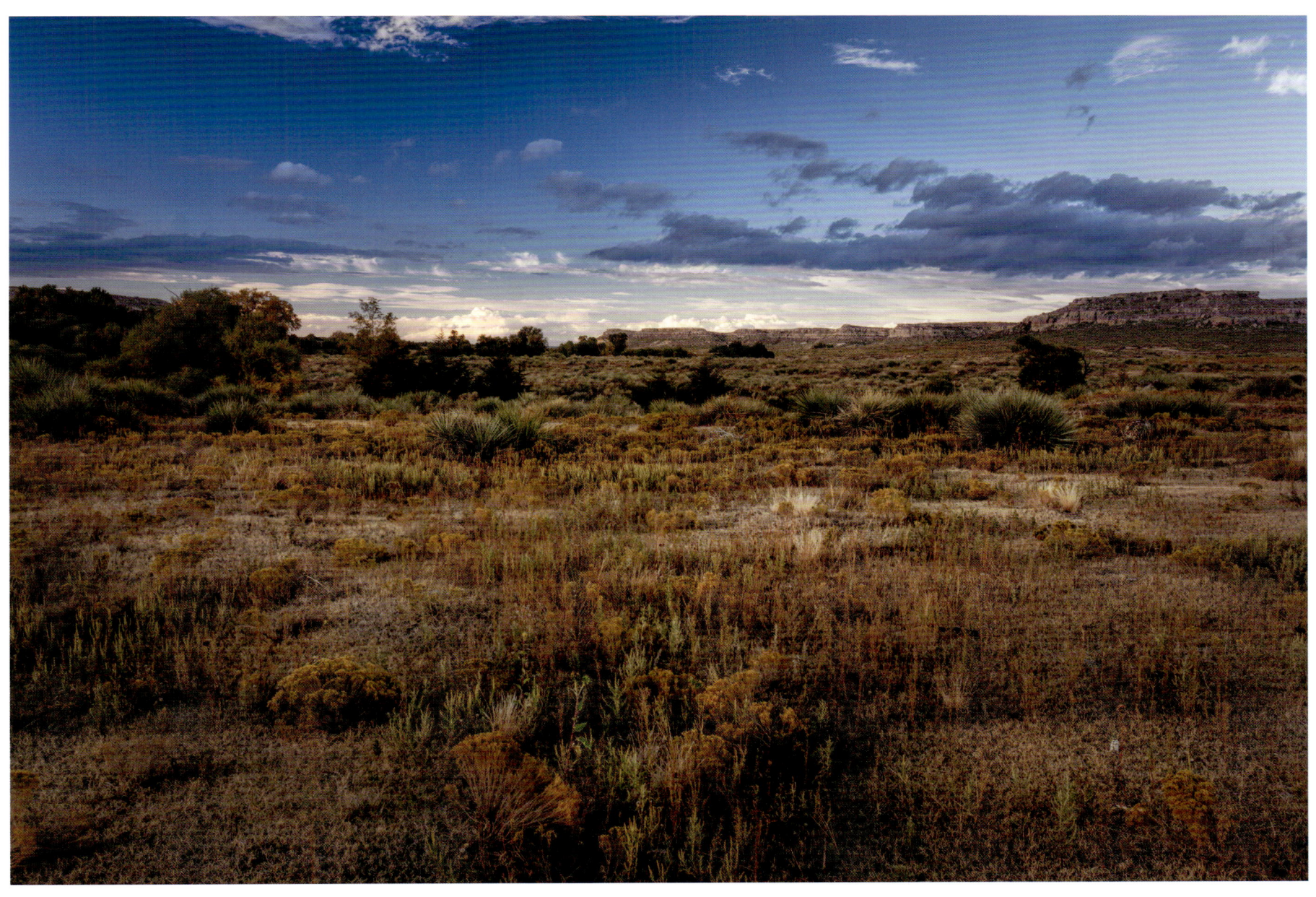

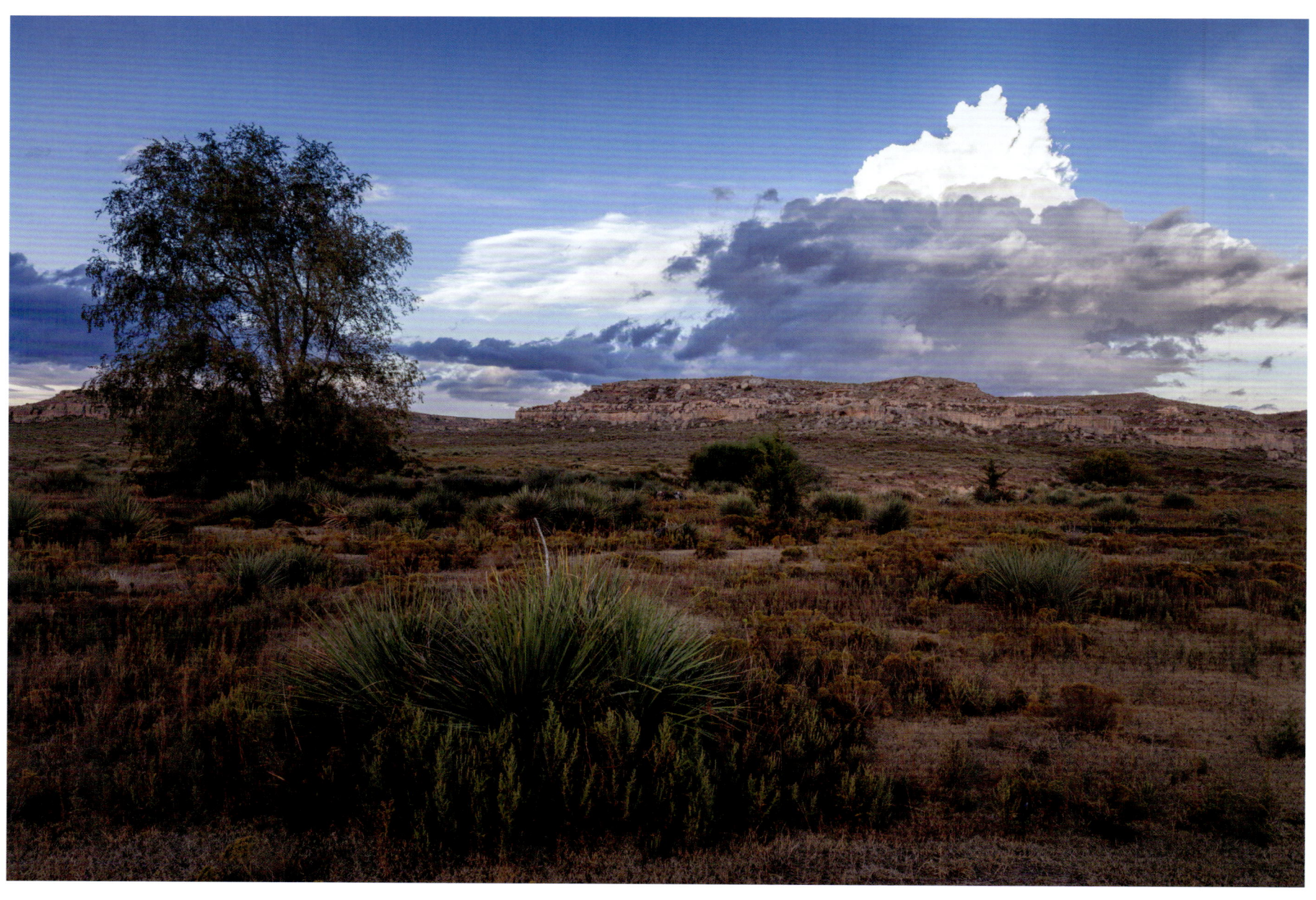

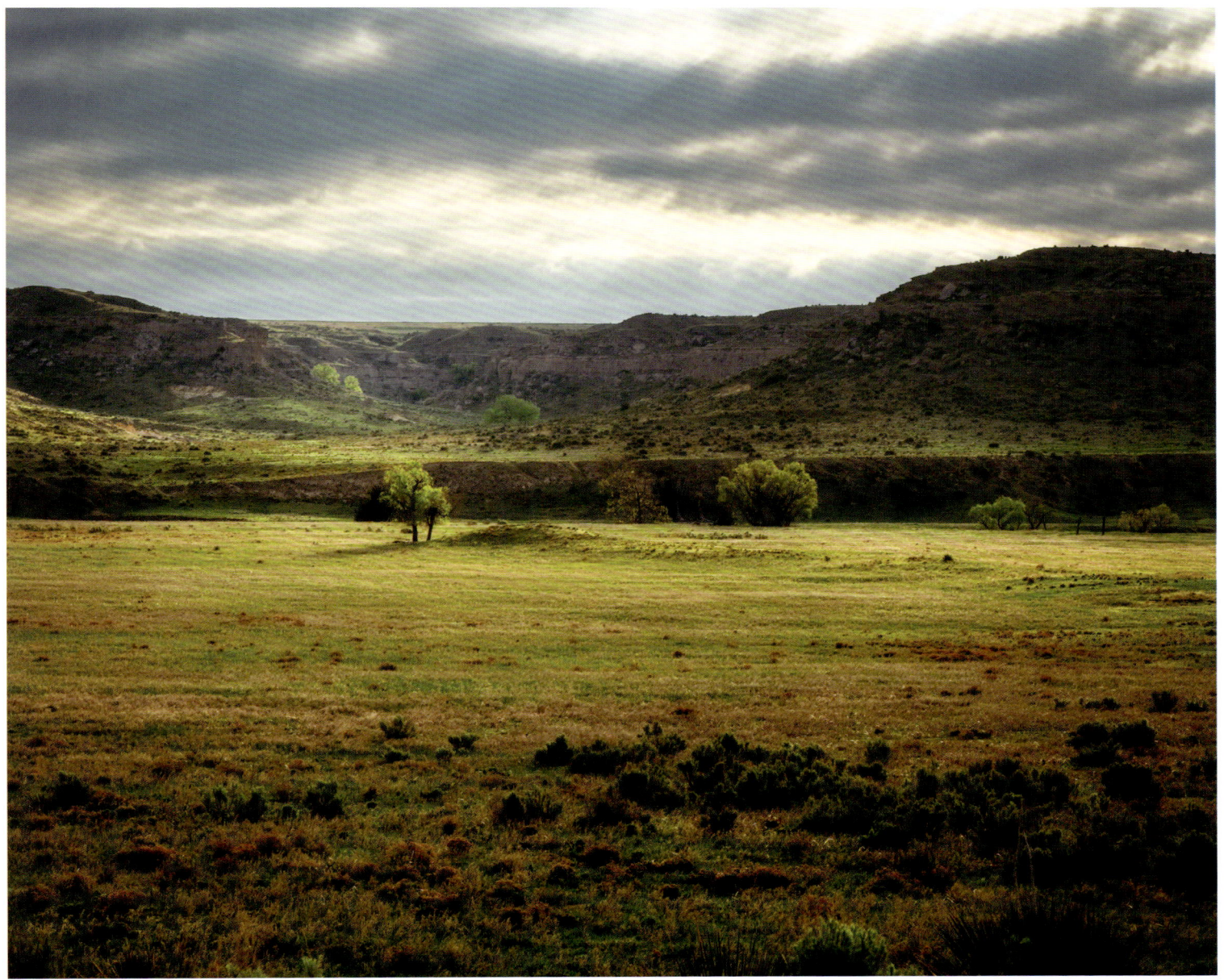

Opposite: Weather is an important element of any landscape, and clouds brought by storms can add visual interest to the sky as well as help shape the mood or feeling of a photograph. A distant storm was developing when I took this photo near Scott County Lake. The storm clouds balanced well with the yucca in the foreground. Yucca are fascinating plants and always provide a nice green contrast to the landscapes in which they are typically found.

Left: The desertlike landscape in Scott County comes to life after an unusually heavy late-spring rain. The flat, stream-carved valley is bounded by steep cliffs on either side, then slowly opens up to the flat plains to the north. (Photo by Mark Clarke.)

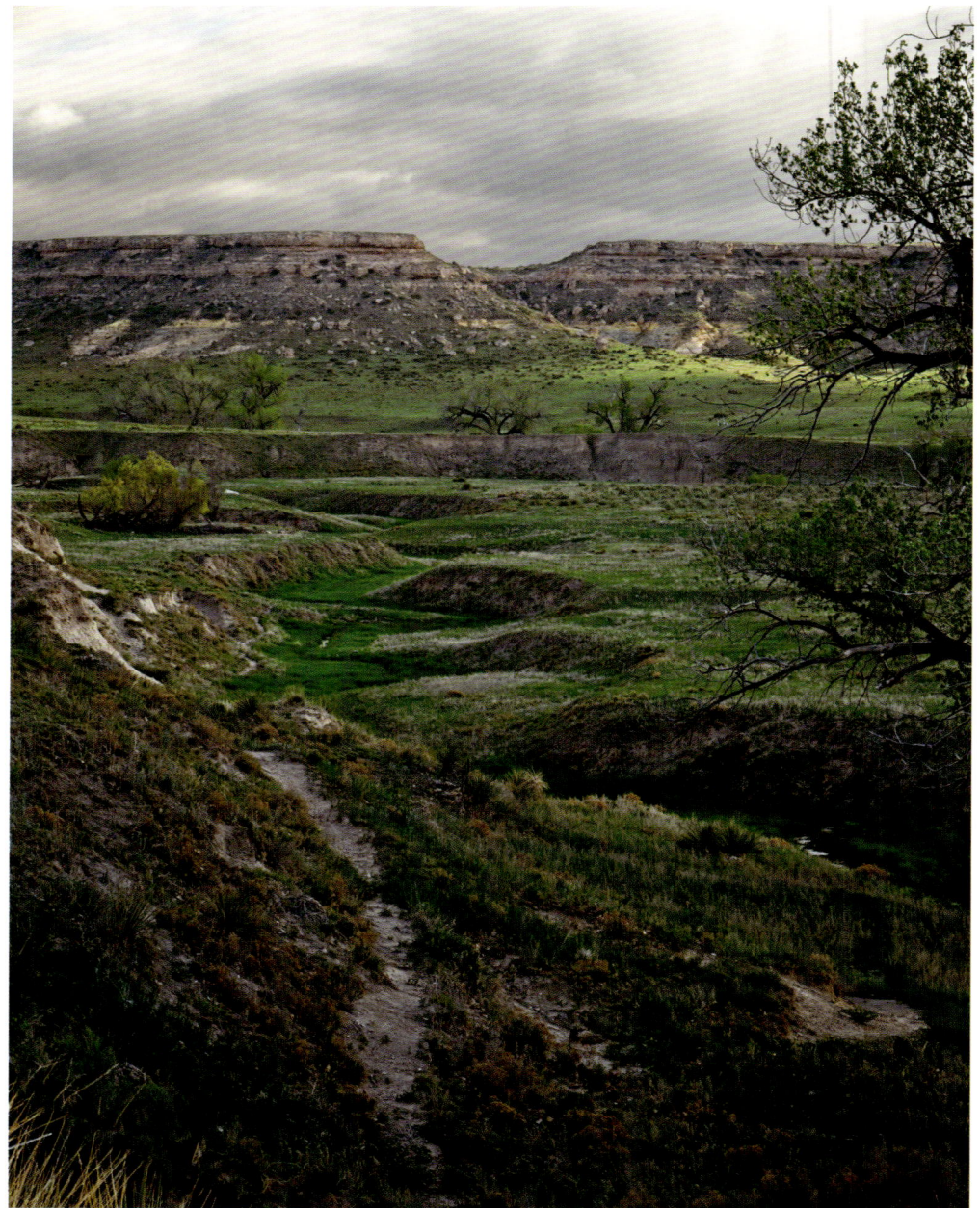

resistant caliche layers. The flat-topped mesas and steep bluffs are typical of semiarid landscapes. These photos also show the characteristic shortgrass prairie, with scattered yucca and prickly pear cactus. Sadly, the shortgrass ecosystem has been under threat from habitat destruction, extermination of native herbivores and predators, and the spread of invasive plants.

In the Arikaree Breaks of Cheyenne County, you can see exposures of the Pleistocene loess that covers much of western Kansas. A narrow area of rugged terrain has ravines and canyons cut into deposits of loess up to one hundred feet thick. The loess was created by the grinding and pulverizing action of the continental glaciers to the north and northeast and was then transported by wind from the Platte River Valley in Nebraska into western Kansas. The steep-sided ravines and canyons were then cut into this loess deposit thousands of years ago by streams during a time of wetter climate. Some of the canyons cut entirely through the loess and into the underlying Ogallala Formation. Despite the fine-grained texture of loess, it has held steep, nearly vertical slopes over the intervening years because the tightly packed angular silt grains are able to provide sufficient cohesion to resist slope failure. The ability of loess to maintain vertical cliffs can also be seen in the photograph of the loess butte.

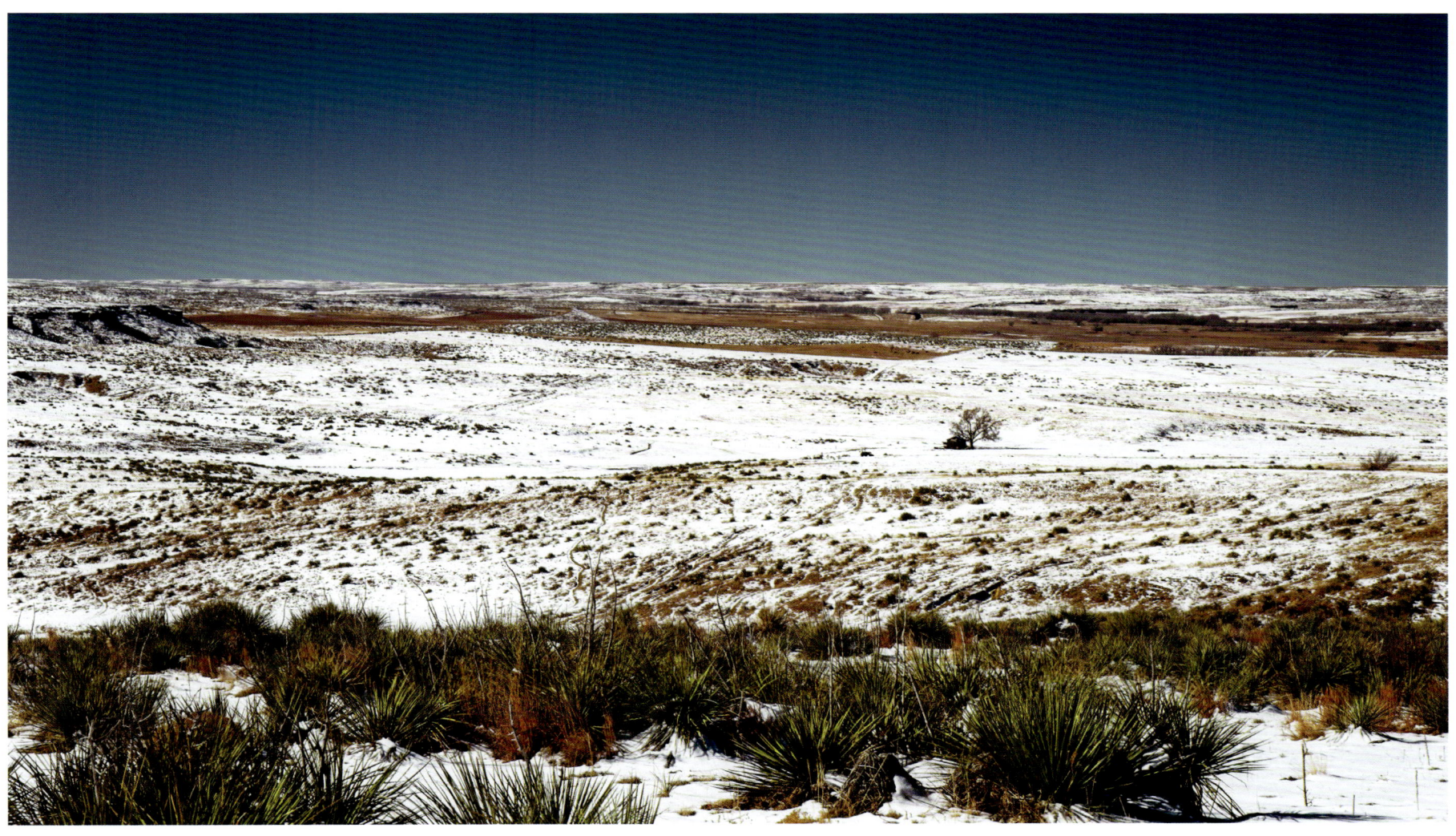

The Arikaree Breaks in far northwest Kansas are a dramatic series of ravines and gullies and are prized mainly for cattle and bison grazing. There are signs of attempts to live within the breaks, as can be seen in this image, with a small abandoned house and single tree in an otherwise empty landscape. (Photo by Mark Clarke.)

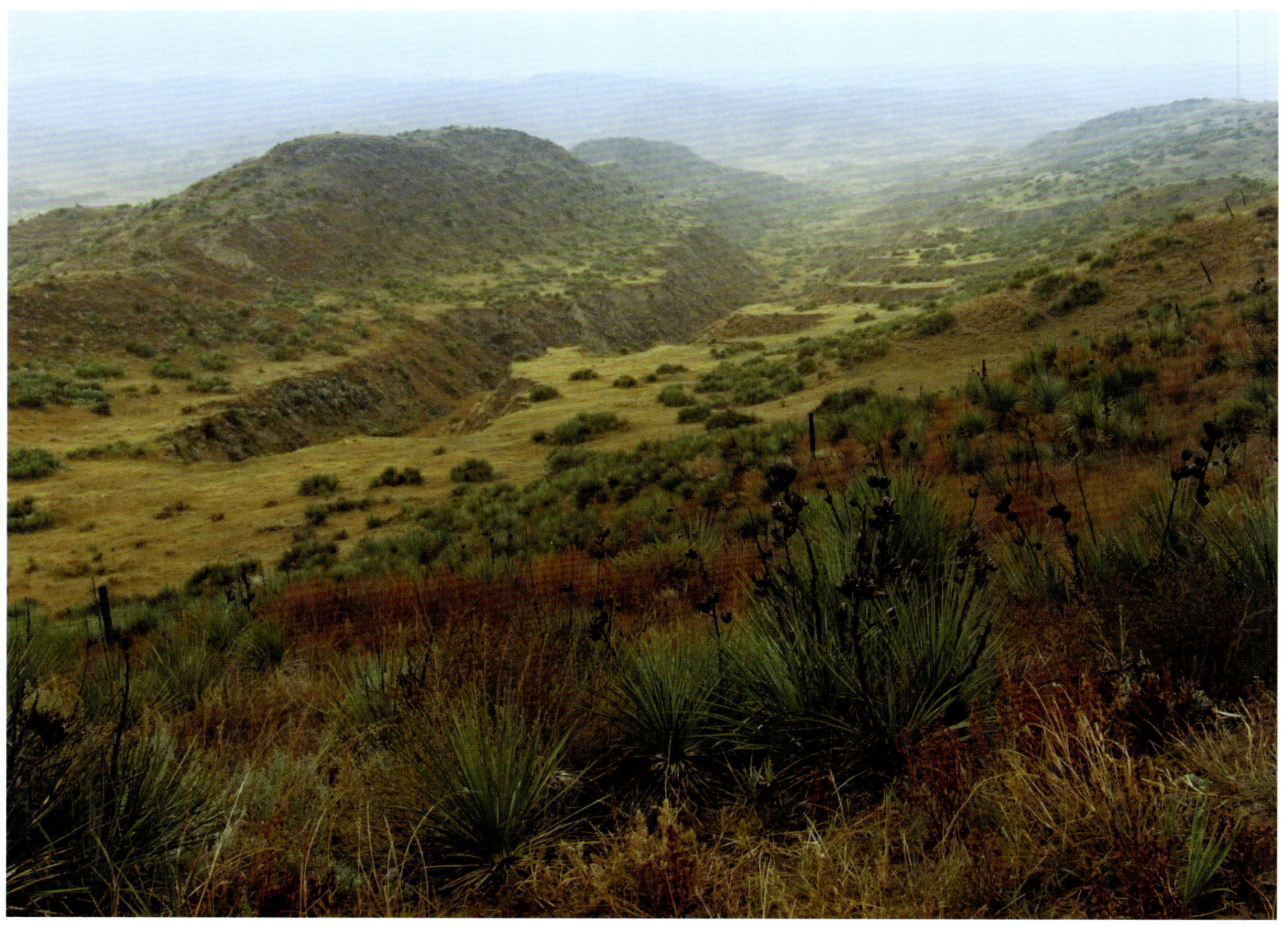

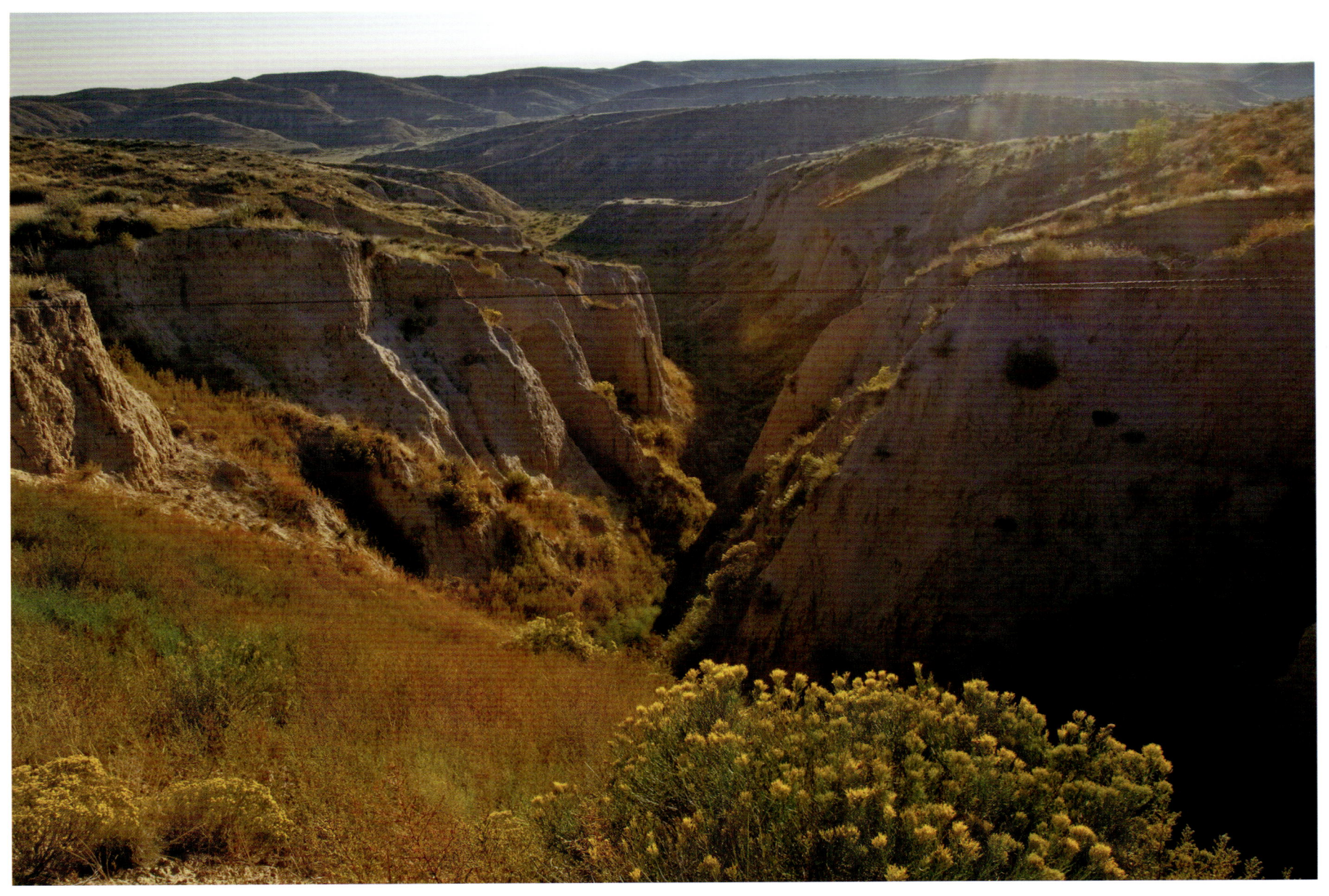

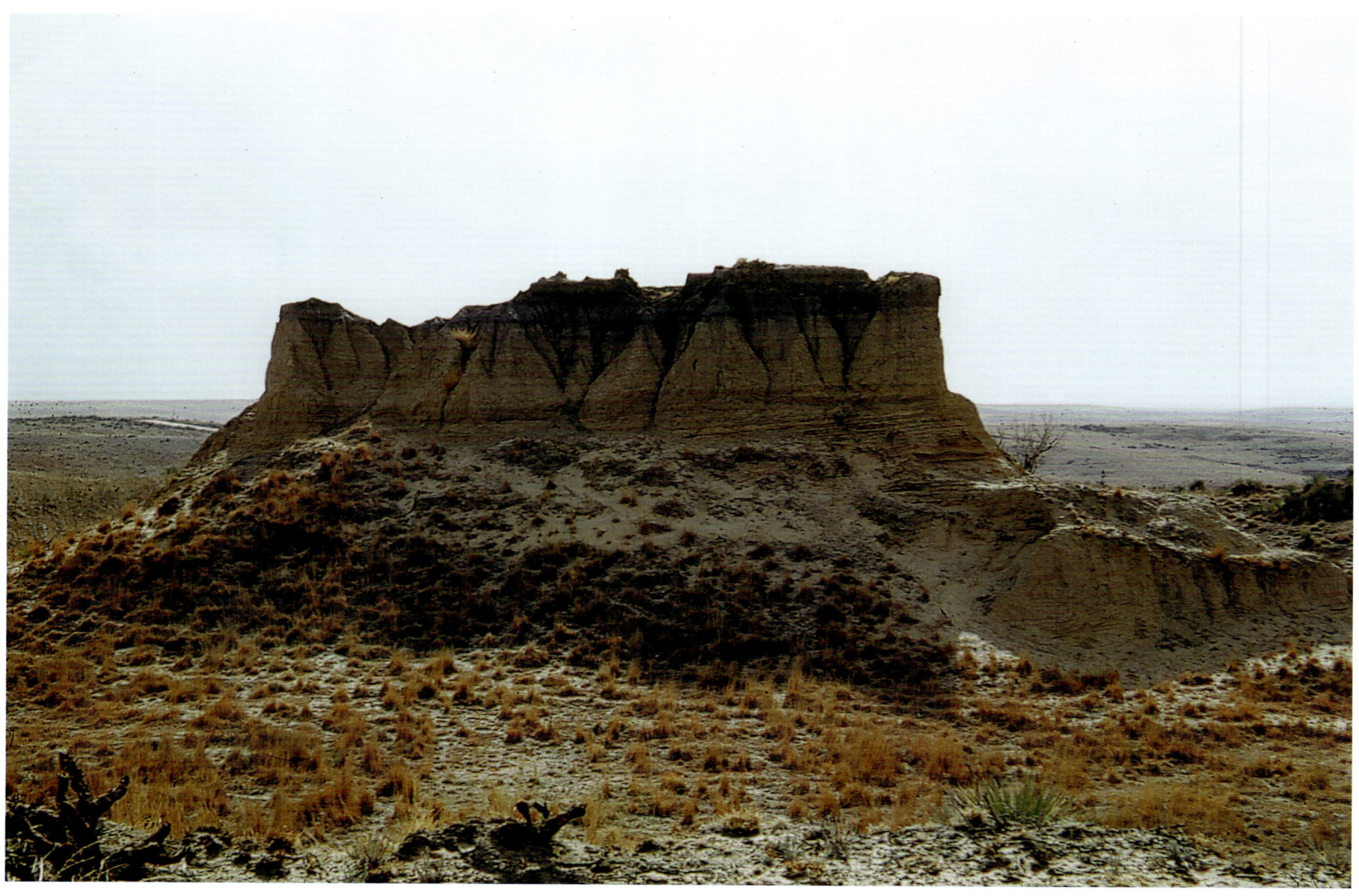

Page 98: One of the most unique landscapes to be found in Kansas is the Arikaree Breaks, which runs through Cheyenne and Rawlins Counties in Kansas and into eastern Colorado. The breaks are a series of deep ravines and gullies, dug through time into the otherwise flat, farmable land around it. The breaks are a hidden gem, and visiting has been on my must-do list for quite some time. (Photo by Mark Clarke.)

Page 99: A typical view of the Arikaree Breaks shot early in the morning in late September. Canyons like these dominate the landscape in this extreme northwest part of Kansas. (Photo by Mickey Shannon.)

Opposite: A loess butte in Cheyenne County is a remnant of the thick layer of loess that once covered extreme northwestern Kansas. (Photo courtesy of the Kansas Geological Survey, Grace Muilenburg, photographer.)

Section 7:
Glaciated Regions

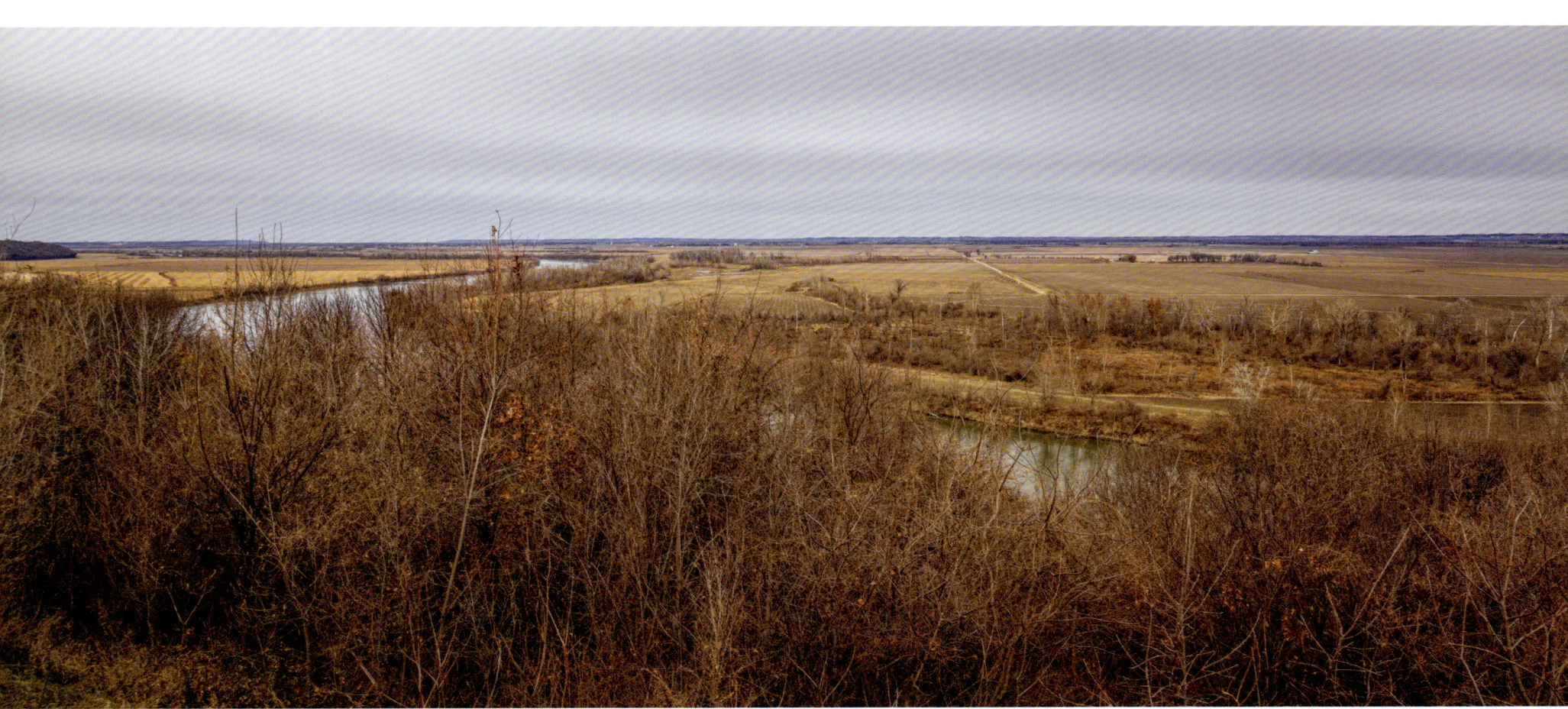

GLACIATED REGIONS: BURIED UNDER ICE

Remnants of the Ice Age

During the Pleistocene Epoch, between 2.6 million and 11,700 years ago, continental ice sheets advanced and retreated multiple times across parts of the northern United States. As glaciers advanced and retreated, the global sea level rose and fell, impacting coastlines and changing the behavior of rivers. However, unlike during the Pennsylvanian/Permian ice age, the midcontinent was never flooded because the continental interior had already been uplifted several thousand feet.

The Pleistocene ice sheets twice reached into northeastern Kansas, first reaching just the far northeastern corner of the state and the second time reaching as far as Topeka, Lawrence, and Kansas City along the present course of the Kansas River. This second glacial advance occurred about 600,000 years ago, and ice was up to five hundred feet thick over northeastern Kansas. When the ice melted, deposits of ice-transported and -deposited sediment were left behind. These deposits, called glacial drift, are a chaotic mixture of silt, sand, pebbles, and boulders. Although most glacial landforms, such as moraines, were subsequently removed by erosion, large boulders still cover some hillsides. These boulders were carried and dropped by the glacial ice after being transported hundreds of miles from their original source. Because these boulders have compositions different from those of any other rocks in the area where they are found, they are called "erratics." The most abundant erratics in Kansas are of Sioux Quartzite, consisting of pink metamorphosed sandstones and conglomerates over a billion years old that were carried southward from southern Minnesota, South Dakota, and northwestern Iowa.

Beyond the Ice Margins

Glacial ice and sediments dammed rivers and changed drainage patterns. The advancing ice eroded south-trending valleys, and glacial drift buried and filled others that flowed eastward. These drift-filled valleys can be found throughout the Glaciated Region and beyond the edges of the ice sheets. Before glaciation, the ancestral Missouri River was a tributary to a larger Kansas River, but the rearrangement of drainage patterns by the advancing ice made the Missouri the major channel for meltwater following glacial retreat. During maximum glacial advance, ice lobes dammed the Kansas River valley. As the ice melted, the Kansas River valley became a pathway for the flow of glacial meltwater that filled the valley with glacial sediments.

Meltwaters from the ice sheets flowed over outwash plains, transporting and reworking the glacial sediments and forming temporary glacial lakes. When the ice retreated, the silts and fine sands of these lake beds and outwash plains were exposed to strong winds that carried the fine sediments away and deposited them as sand dunes or thick layers of silt called loess. Loess hills were deposited along the margins of the Missouri River in Kansas, Missouri,

Page 103: This view over the Missouri River looks to the northeast from the high loess bluffs at the White Cloud "Four State Lookout" overlook in Doniphan County. It is hard to imagine that this entire view would have been overridden by continental glaciers about 600,000 years ago.

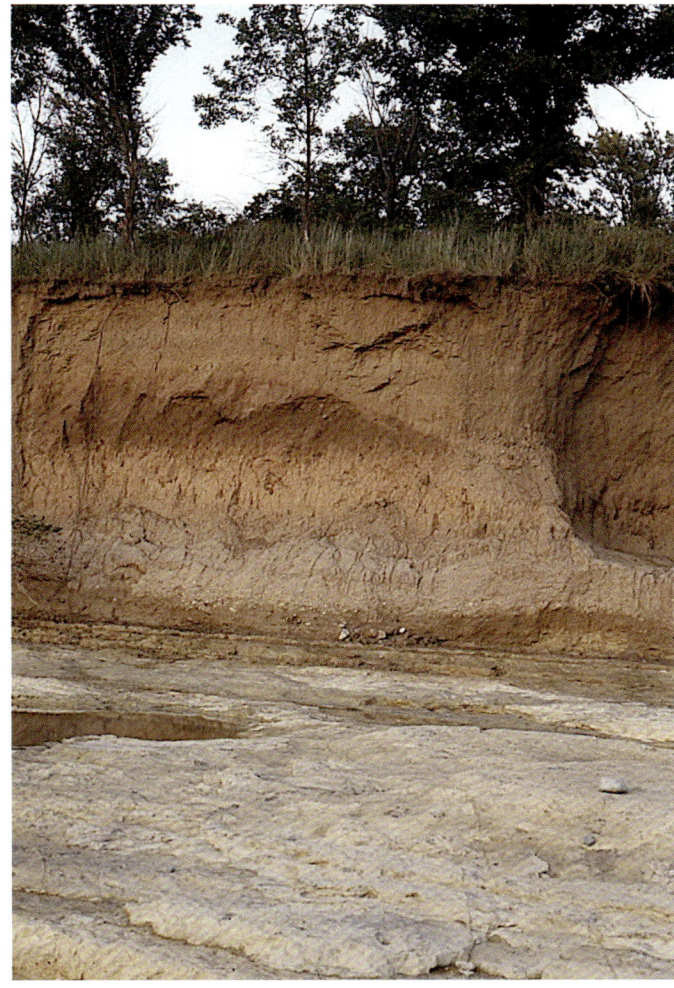

This thick layer of windblown silt to fine sand was deposited during a time of Pleistocene glaciation and rests on a 300-million-year-old Permian limestone. These loess deposits often preserve horizons of soil development representing Pleistocene land surfaces and contain a variety of mammal bones. (Bank of the Big Blue River south of Tuttle Creek Reservoir in Riley County. Photo by Keith Miller.)

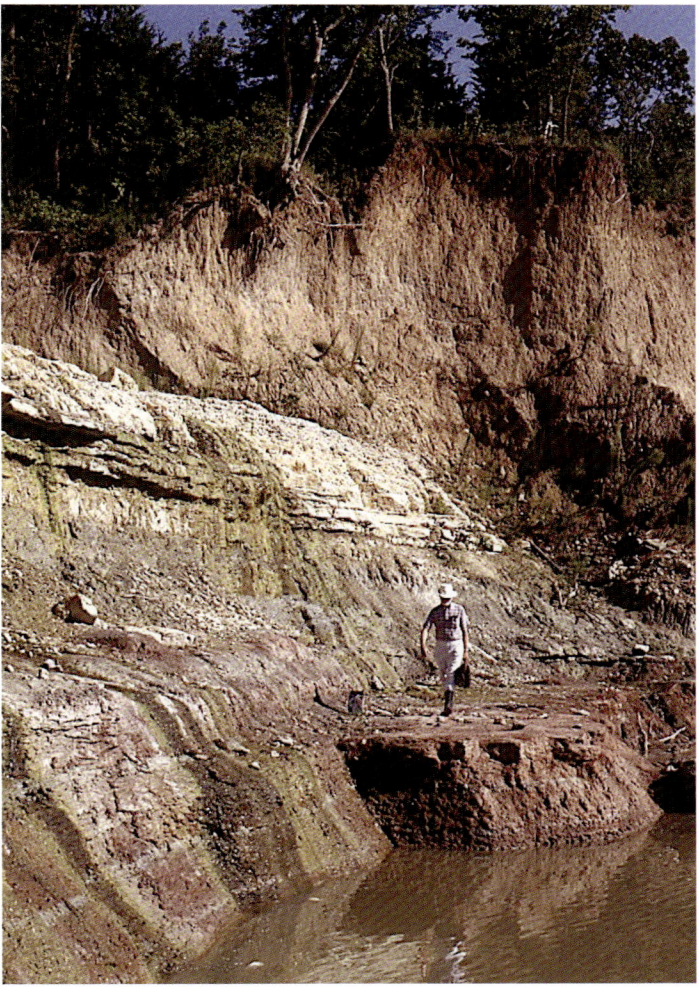

A small valley eroded into early Permian rocks of the upper Blue Springs Shale and lower Florence Limestone was filled by loess blown from the margin of an advancing glacier. The wall of this ancient buried valley can be seen cutting diagonally across the image. (Pleistocene valley fill exposed by recent flood erosion in the Milford Spillway near Junction City, Geary County. Photo by Keith Miller.)

and Iowa. Loess deposits can also be seen overlying bedrock, particularly on the eroded banks of stream and river valleys in eastern Kansas. Windblown silt can be transported very long distances, and the loess mantling much of western Kansas in the High Plains was carried all the way from the Platte River Valley in Nebraska. Unlike the fine silt of loess, sand is moved near ground level by wind and deposited relatively near its source as migrating sand dunes. Extensive dune sands are found within the Arkansas River Lowlands, having been derived from sediments transported by the river channel.

As the continental glaciers repeatedly advanced and retreated, plant and animal communities migrated back and forth. As a result, Pleistocene fossil plant and animal species in Kansas represent both cold glacier-margin climates and warmer interglacial climates. The glacial deposits of Kansas contain abundant fossil remains of the vertebrate communities that inhabited Kansas during the Pleistocene. Pleistocene fossils regularly erode from the unconsolidated glacial sediments at the banks of streams. Mammal fossils include the bones and teeth of mammoths, mastodons, bison, camels, horses, llamas, ground sloths, and dire wolves as well as saber-toothed cats. Lastly, near the surface one can find the record of North America's native peoples who migrated to the great plains hunting large mammals during the closing millennia of the Pleistocene as the last glaciers retreated.

Glaciated Regions Landscapes

The Missouri and Kansas Rivers and their tributaries were extensively modified as the continental glaciers advanced southward. Depending on their direction of flow, river and stream courses were eroded, dammed, filled by glacial sediment, or diverted.

The Missouri River valley was covered in ice during the advance of continental glaciers into northeastern Kansas. When the glaciers retreated, the Missouri became a major pathway for huge volumes of meltwater. Consequently, large amounts of glacial sediment were deposited within the Missouri flood-plain. The Wolf River, a small tributary to the Missouri, flows eastward and then turns northward to flow into the Missouri River. Similar to the Missouri, the area of the Wolf River would have been entirely overridden by glacial ice.

Wakarusa Creek, like the Kansas River just to the north, flows from west to east. At the greatest extent of ice advance, the Kansas River valley was filled by ice. With the Kansas blocked, the Wakarusa swelled with meltwater, resulting in increased erosion and sediment transport. Lakes formed on tributaries, and some of these overflowed stream divides. As the ice retreated, water began flowing again within the larger valley of the Kansas River, causing it to erode downward. In response to this lowering of the Kansas valley, the Wakarusa cut downward to keep pace. Stream valleys throughout the Glaciated Region have similar complex histories.

The loess hills along the Missouri River valley were formed during the Pleistocene by strong winds carrying fine glacial outwash sediment away from the river floodplain. This fine sediment was deposited as irregular, hummocky hills in northeastern Kansas that were later stabilized by vegetation during wetter postglacial climates. Despite their fine-grained composition, the loess hills have sharp crests and sometimes very steep slopes. The Glacial Hills Scenic Byway extends sixty-three miles along Highway 7 and provides views of the Missouri River and loess hills from Leavenworth in the south to White Cloud at the Nebraska border. The byway has been designated by the National Park Service as part of the official route of the Lewis and Clark National Historic Trail.

Scattered over the landscape of northeastern Kansas are erratic boulders that were carried southward hundreds of miles by the glaciers and then dropped

I was on a fall colors trip in northeast Kansas, specifically centering on Atchison and Leavenworth. One of my favorite spots on this visit was the vantage point at Benedictine College looking out over the bend of the Missouri River. With a little fog coming up off the river on a crisp autumn morning, it was a perfect scene as the sun warmed the air. (Photo by Mickey Shannon.)

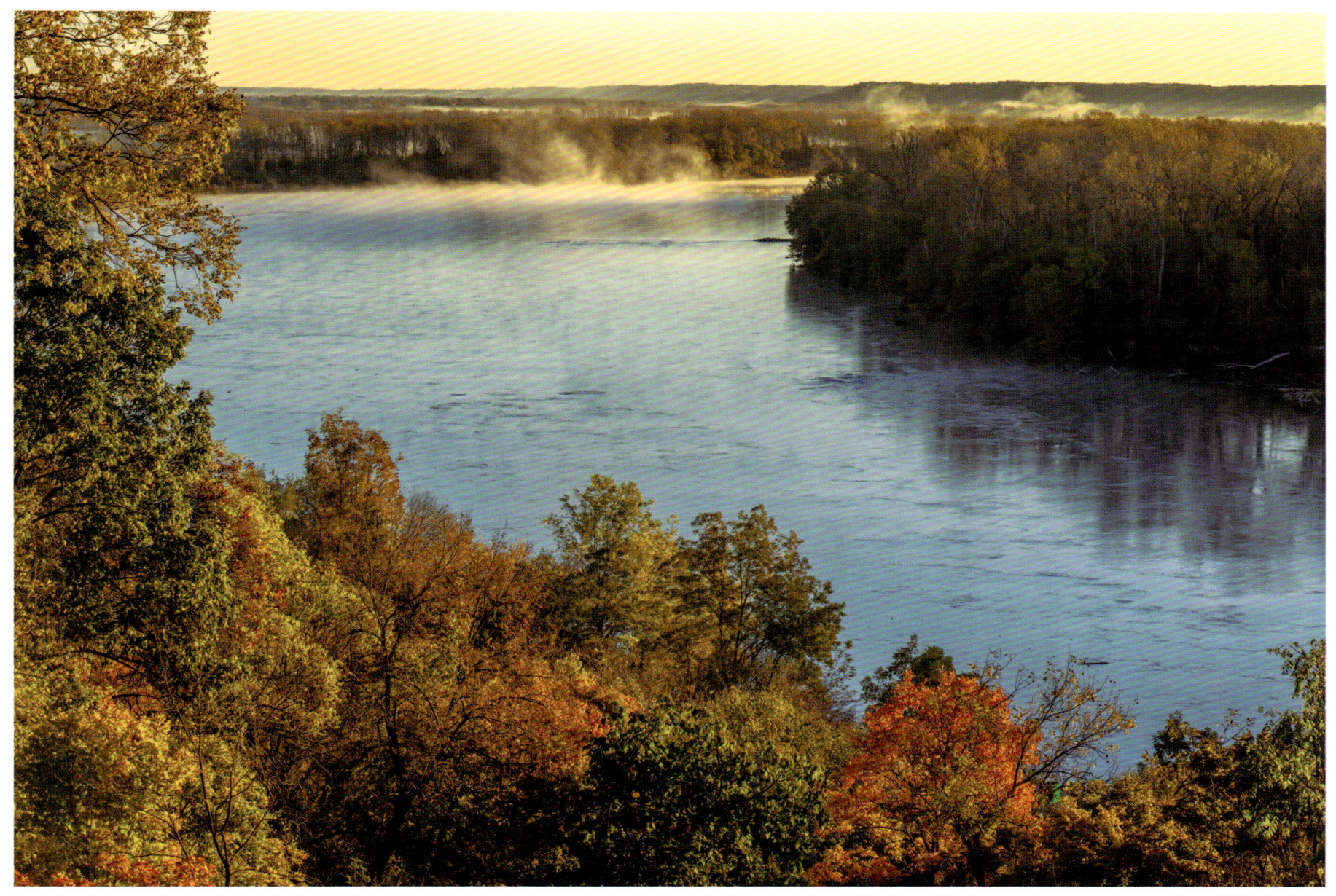

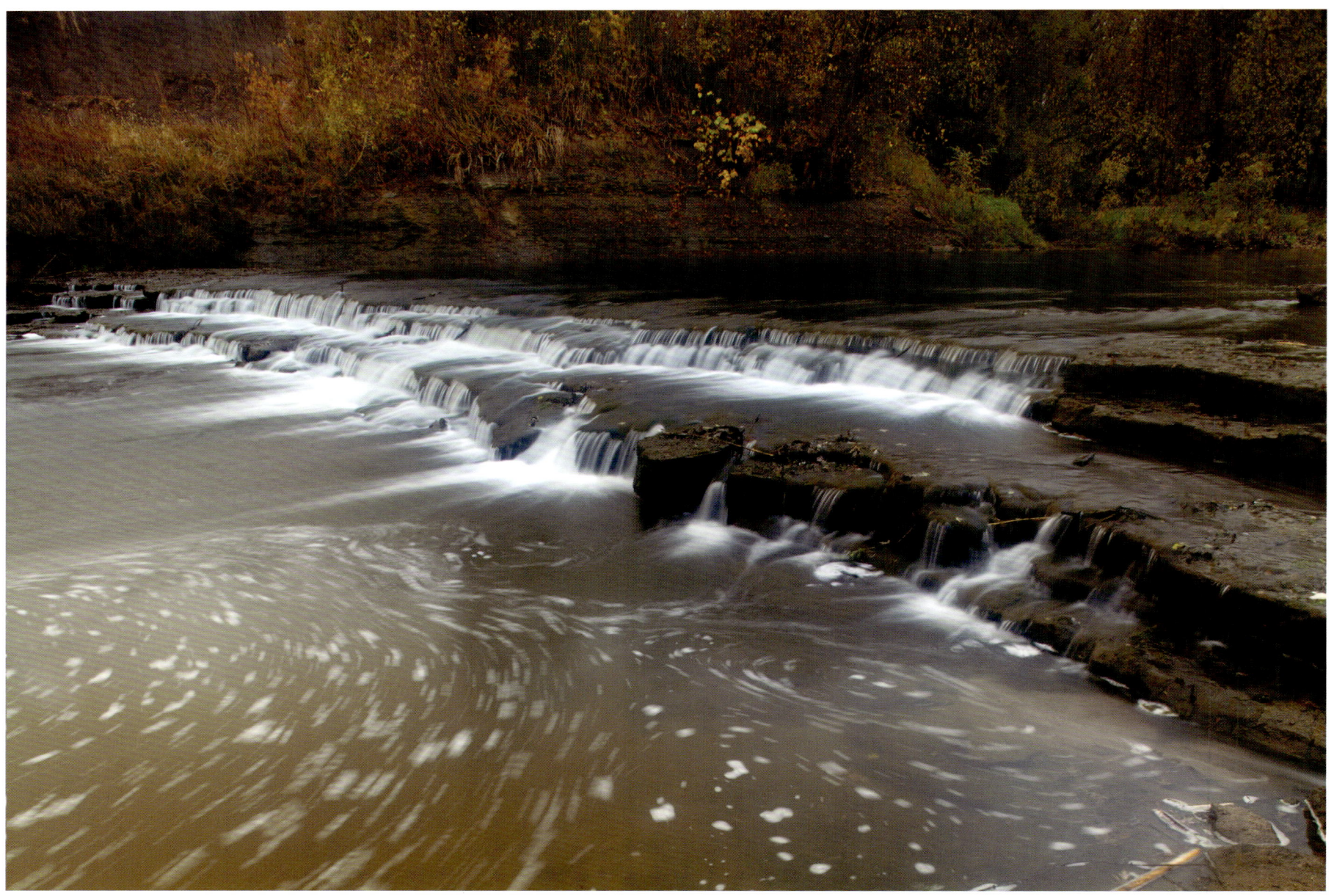

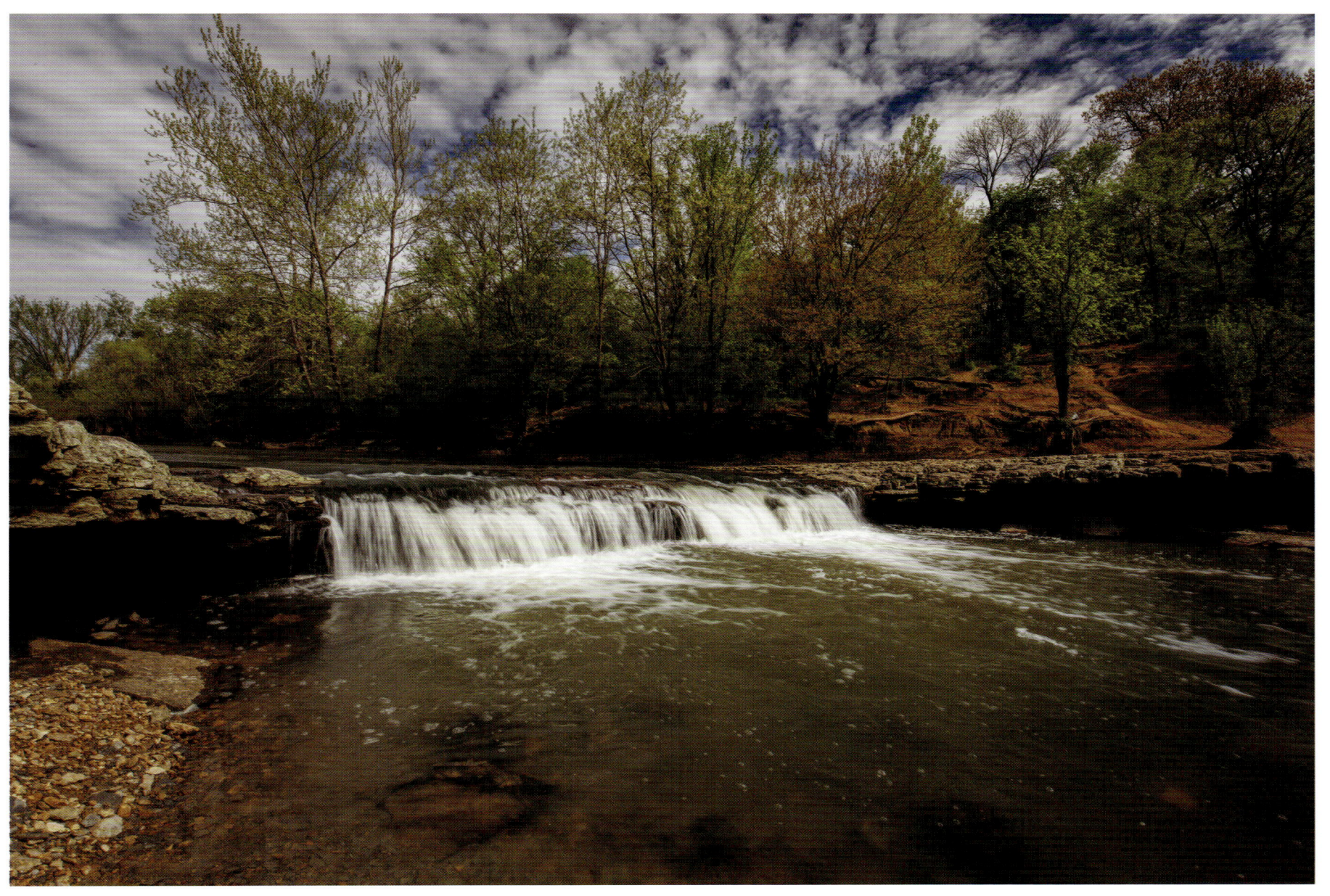

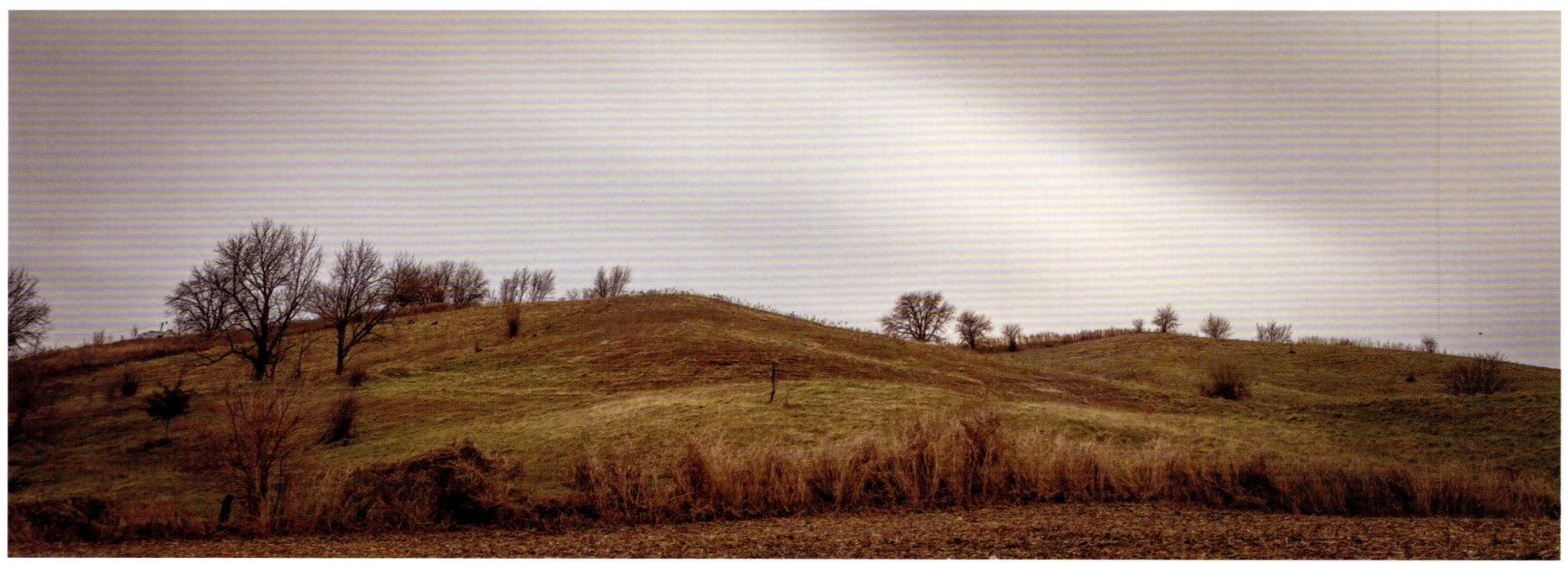

Page 108: This photo was taken near the Glacial Hills Scenic Byway in northeast Kansas, showing the second of three sets of waterfalls along the Wolf River near Severance in Doniphan County. This was shot on one of my earliest waterfall trips to northeast Kansas. It was the last photo of the day. It was a horrible rainy day when I hiked out to check this waterfall. I was happy to get back to my car with a shot of it, as I had researched the location for some time. (Photo by Mickey Shannon.)

Page 108: A perfect spring morning for a visit to Wakarusa Falls in Douglas County near Lawrence. I love the sound of running water, and this was a perfect place to spend a bit of time.

Above: The unique topography of this hillside along the Glacial Hills Scenic Byway in Doniphan County caught my attention as I was driving by. Shapes and textures always initially grab my attention. The repeated triangular patterns in the hills and hillsides provide texture and visual interest.

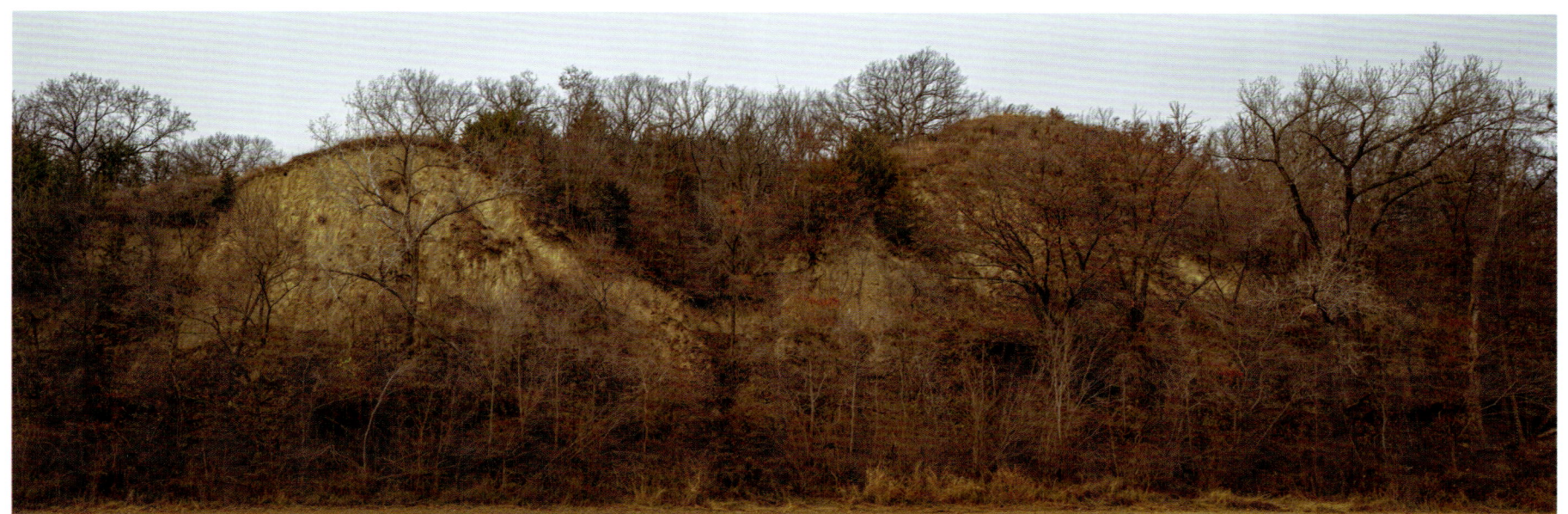

Naturally, the visual aspects of an area are what catch my attention. However, I'm starting to think more about how the landscapes I'm photographing were shaped. What geological processes did the area go through? Over what expanse of time was the landscape formed? Thinking about everything that had to happen to create these loess hills in Doniphan County gives me a new perspective as I view these landscapes.

as the ice melted. The most common erratics are composed of Sioux Quartzite, a metamorphic quartz sandstone that is cemented by silica and is extremely hard and resistant to breakage. The erratics range in size from cobbles to boulders the size of cars and weighing ten tons or more. Sioux Quartzite erratics have been used throughout northeastern Kansas for stone buildings, stone walls, and community monuments. The Sioux Quartzite bedrock, from which those erratic boulders came, can be seen today at Sioux Falls, South Dakota. Less common erratics include granite, agate, and Precambrian volcanic rocks from the Lake Superior region. By retracing the places of origin of these ice-transported rocks, geologists can reconstruct the southward movement of the continental glaciers.

Glacial erratics can be seen in the beds of streams in the Glaciated Region where they have been moved downstream during times of high water. Sioux Quartzite boulders have also been left stranded on hilltops in places such as Wabaunsee County where modern weathering has had little impact on the hard, resistant quartzite and erosion has been unable to move them.

I've always found water peaceful, either for the sound of it flowing or its stillness, as in this scene in Wabaunsee County. Just sitting on the glacial rocks by the side of the stream, you can enjoy a few peaceful moments.

Opposite: This quartzite boulder in Wabaunsee County is just a rock on the prairie, but imagine the stories this rock could tell . . . and will still be telling long after I'm gone.

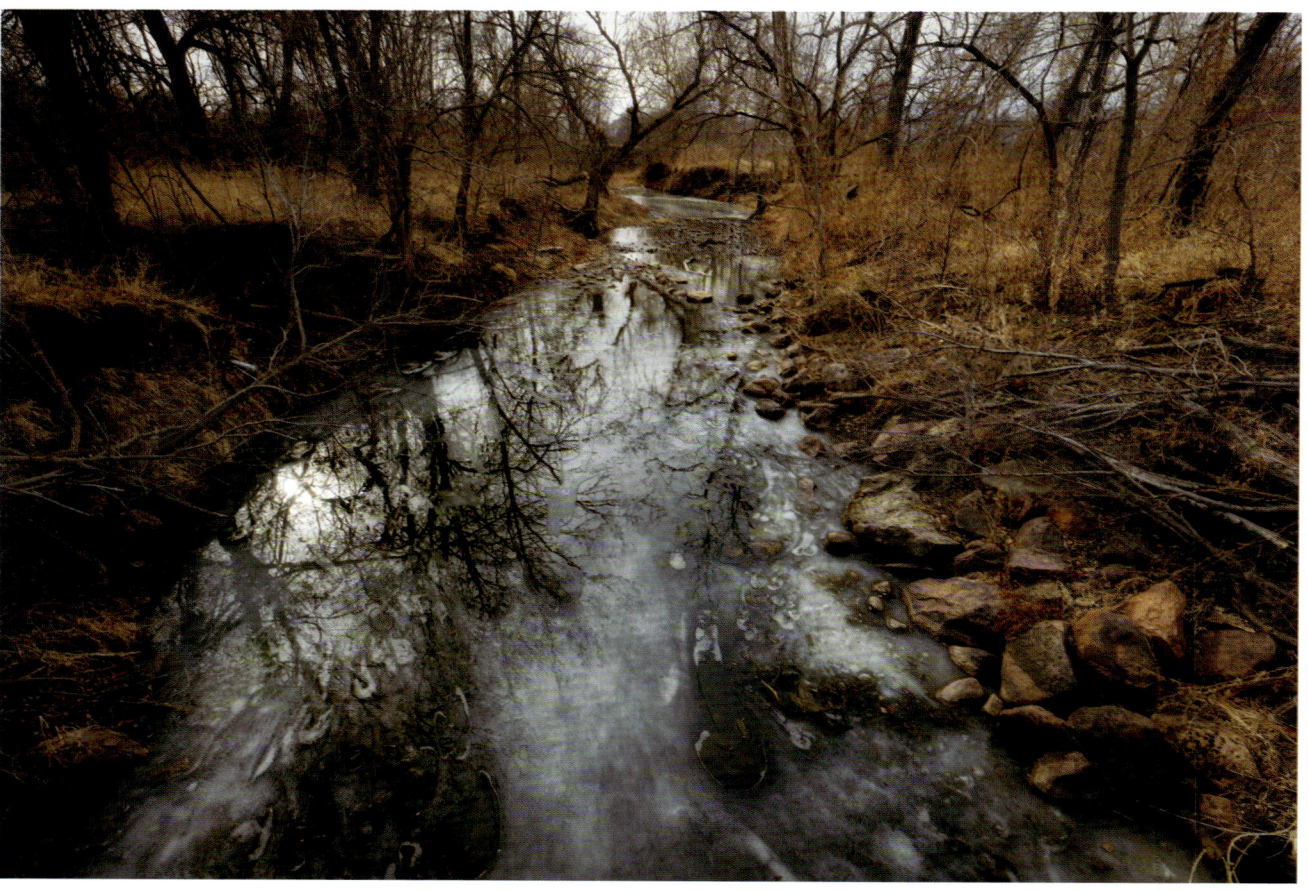

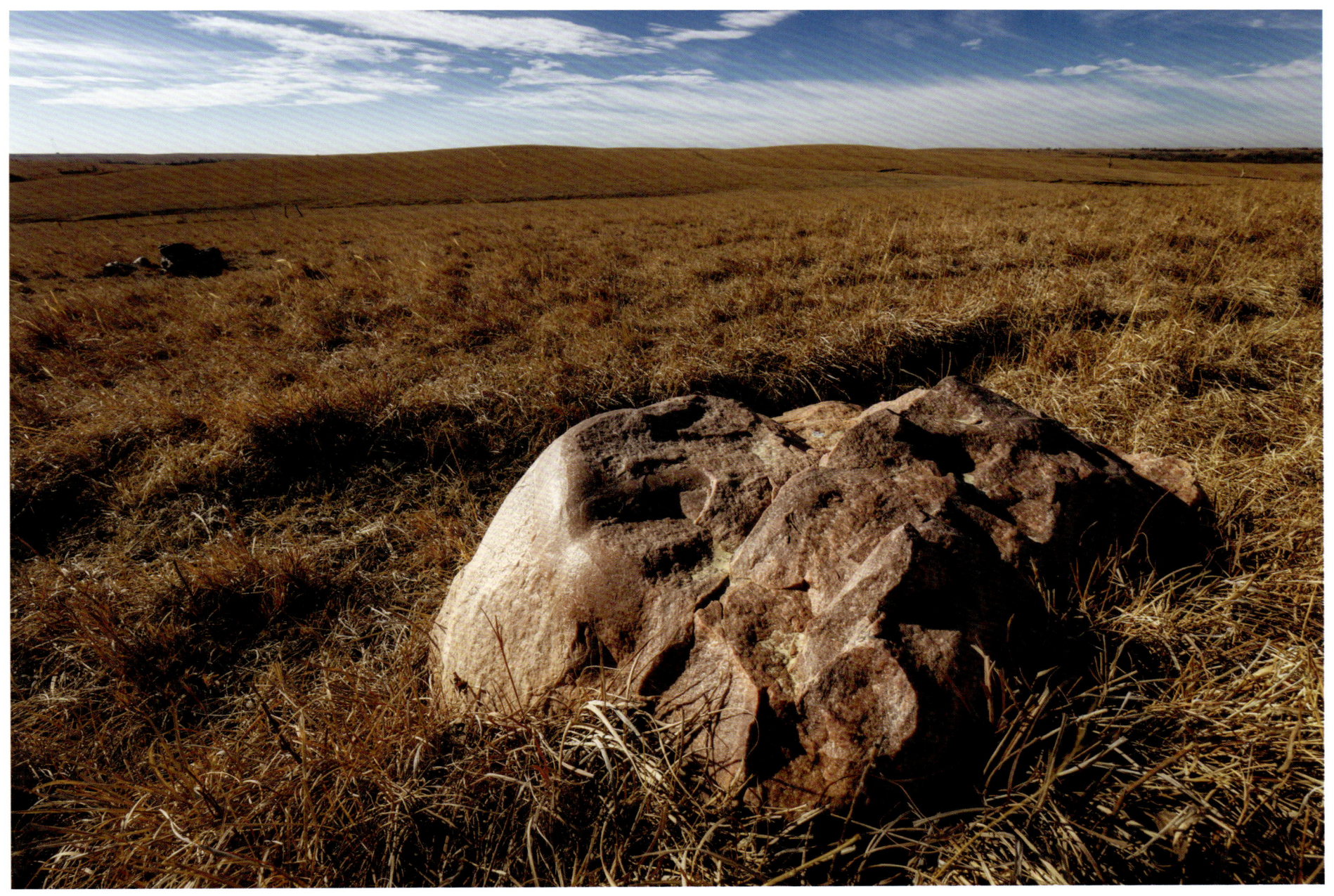

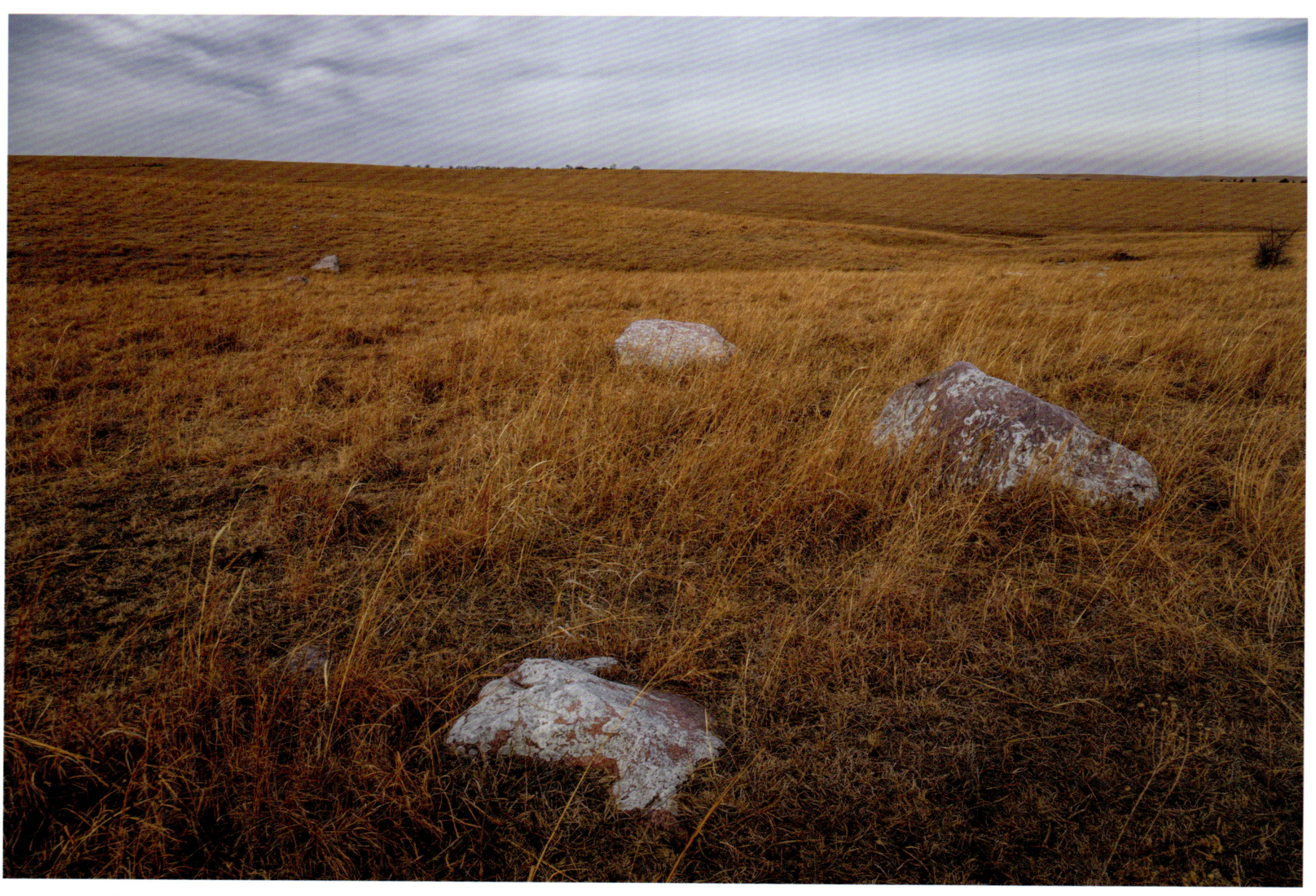

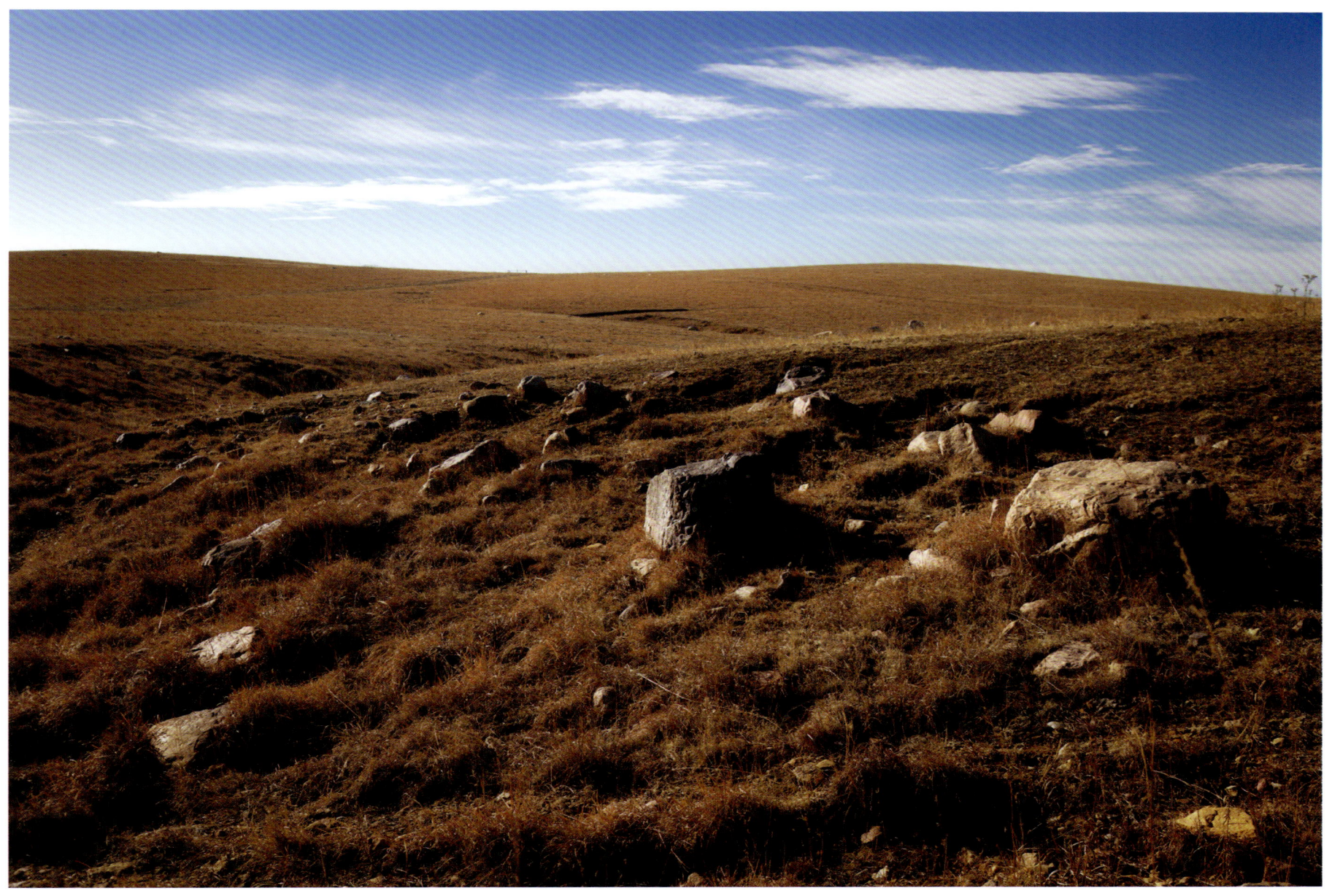

Page 114: It is fascinating to think about everything that had to happen for these glacial erratic rocks to end up where they are on this prairie hilltop in Wabaunsee County. Perhaps there is a parallel with everything that had to happen for me to end up photographing them on the day I took this photo.

Page 115: I like to find views that I can return to in different seasons. The colors of the prairie, the direction of the light, and the look and feel of the landscape will change. This photo is from a view I recently found in Wabaunsee County, one that I will visit again to see what the changing seasons reveal.

Section 8: Arkansas River Lowlands

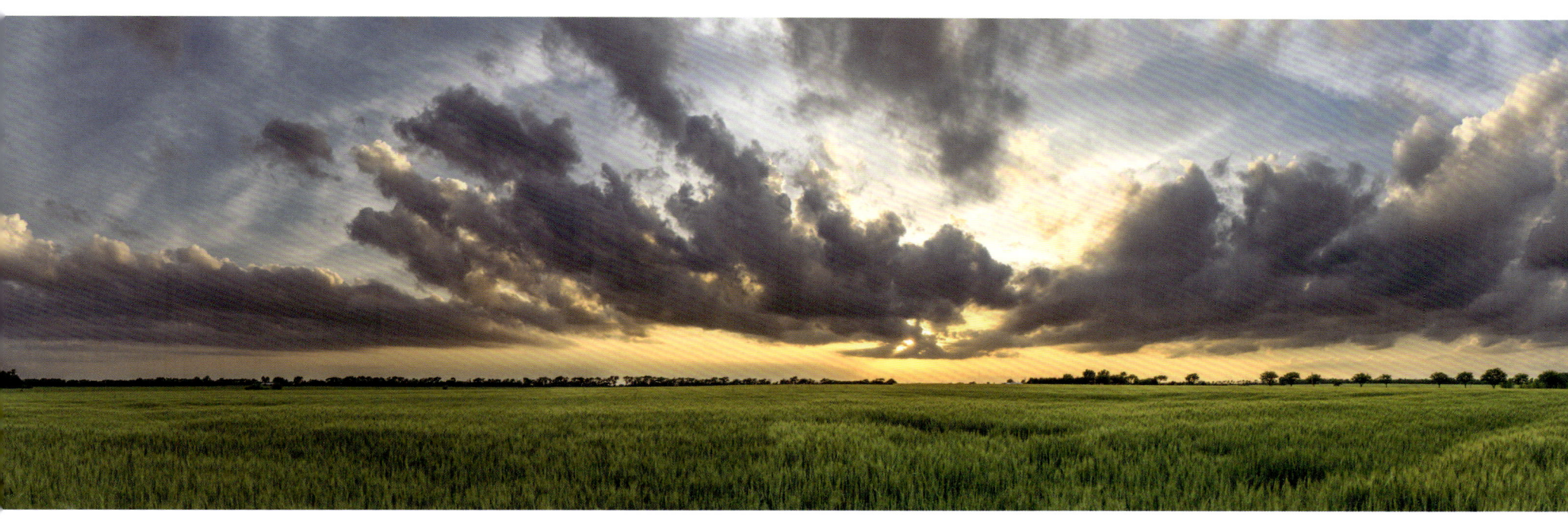

Arkansas River Lowlands: Deposits of a Lazy River

The Arkansas River has flowed from the Rocky Mountains through Kansas for the last ten million years. Over that time, river water from rain and snowmelt has carried sediment eroded from those mountains into the state, depositing it within the migrating channel and floodplain of the Arkansas. The river's unconsolidated deposits of sand, silt, and clay are called "alluvium." Because the land is relatively flat, the river flows slowly and drops most of its sediment onto the Kansas plains. The fertile soils developed on the fine sediment of the floodplains are good for farming.

Because of the low annual rainfall in the High Plains, little water is added to the river as it flows across western Kansas. Furthermore, much of the rain that does fall seeps into the ground or evaporates. Where the groundwater is near the surface, water can flow into the channel. However, as groundwater levels have declined with increased irrigation, parts of the Arkansas River channel in western Kansas are often dry. Because of the high rates of evaporation as the river flows across Colorado and into Kansas, the Arkansas River water in western Kansas is among the saltiest in the country. Some of this highly saline water flows from the channel into the Ogallala Aquifer below. However, in central and eastern Kansas, increased rainfall restores the river flow.

Windblown loess deposits also accumulated on the lowlands during the Pleistocene. In addition, some places south of the river are covered in sand dunes composed of river and glacial sediments blown from the river flood-plain. These dunes are mostly now inactive and covered with grass and other vegetation.

The flat topography has also resulted in wetland areas that are critical for waterfowl and bird species migrating across the midcontinent. Cheyenne Bottoms and Quivira National Wildlife Refuge are the two most important of these stop-over localities.

Arkansas River Lowlands Landscapes

The fine sediments deposited in the Arkansas River floodplain produce deep, fertile soils. Windblown silts deposited during the Pleistocene also fertilize the sediments with needed mineral nutrients. These flat, nutrient-rich river bottomlands are productive areas for grain production, especially for wheat, a major agricultural product of the area.

The images from Edwards and Stafford Counties show the irregular rolling topography produced by vegetated sand dunes. Strong winds during the dry Pleistocene winters picked up sediments from the river valley and deposited them to the south in dune fields. These dunes later became stabilized by vegetation under wetter climatic conditions. During the extended drought of the dust bowl years, these dunes again became active. In the future, similar periods of drought or the loss of the thin vegetation cover could reactivate the dunes and return the area to a landscape of migrating sand dunes.

The flat, poorly drained topography of the Arkansas lowlands has resulted in ecologically important wetlands. Cheyenne Bottoms Wildlife Area, near Great Bend in Barton County, is the largest marsh in the US interior, covering forty-one thousand acres. It has been designated a "Wetland of International Importance" along with the Quivira National Wildlife Refuge in Stafford County. Cheyenne Bottoms is one of the top staging areas for shorebirds and

Page 117: Fields like this are common across Kansas in May. In June, the wheat has turned golden, becoming ready for harvest. While I love the golden color of wheat, I love these green wheat fields in the spring just as much! It gives a bit of a different look. This was shot by luck. I was out scouting and noticed that the clouds were looking interesting. I went to a local wheat field and took this panoramic photo to highlight the beautiful sunset that was unfolding. (Photo by Mickey Shannon.)

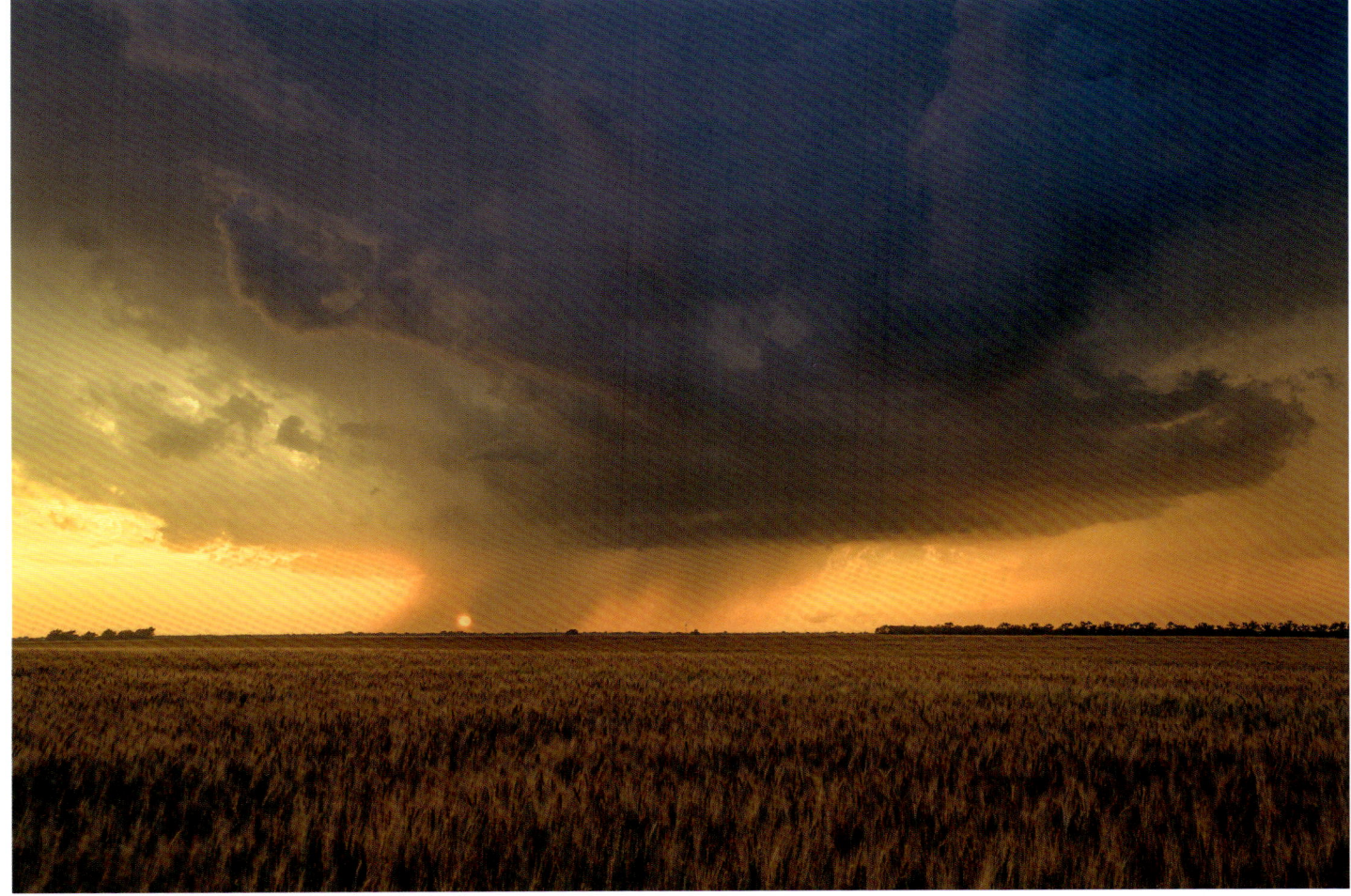

A storm cloud dumps rain during a vivid orange sunset in Sedgwick County. The wheat field seen here was harvested less than a week later. (Photo by Mickey Shannon.)

waterfowl in the United States, with approximately 45 percent of the total North American shorebird population stopping in the area during spring migration. The limited water resources for the marsh are managed with a complex system of dikes and water control structures. Quivira National Wildlife Refuge has over twenty-two thousand acres of inland salt marsh and sand prairie. Both Cheyenne Bottoms and Quivira struggle with maintaining adequate

water for migrating birds, and both are critical habitats for the endangered whooping crane population and other threatened species.

These two internationally significant wildlife habitats, together with the Konza Prairie, the Tallgrass Prairie National Preserve, wooded riparian areas, and areas of preserved mixed-grass and shortgrass prairie, illustrate the ecological diversity and biological importance of Kansas landscapes.

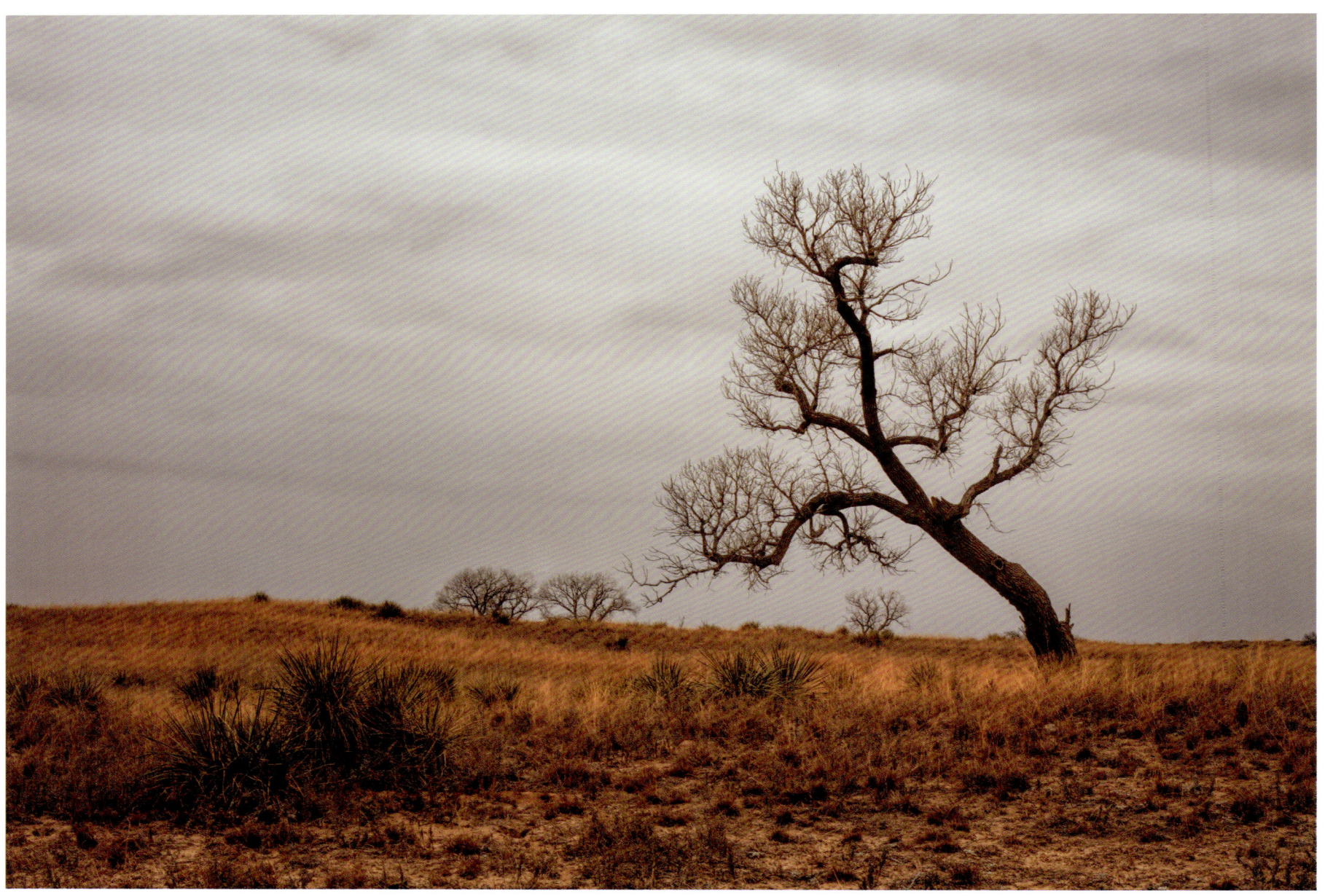

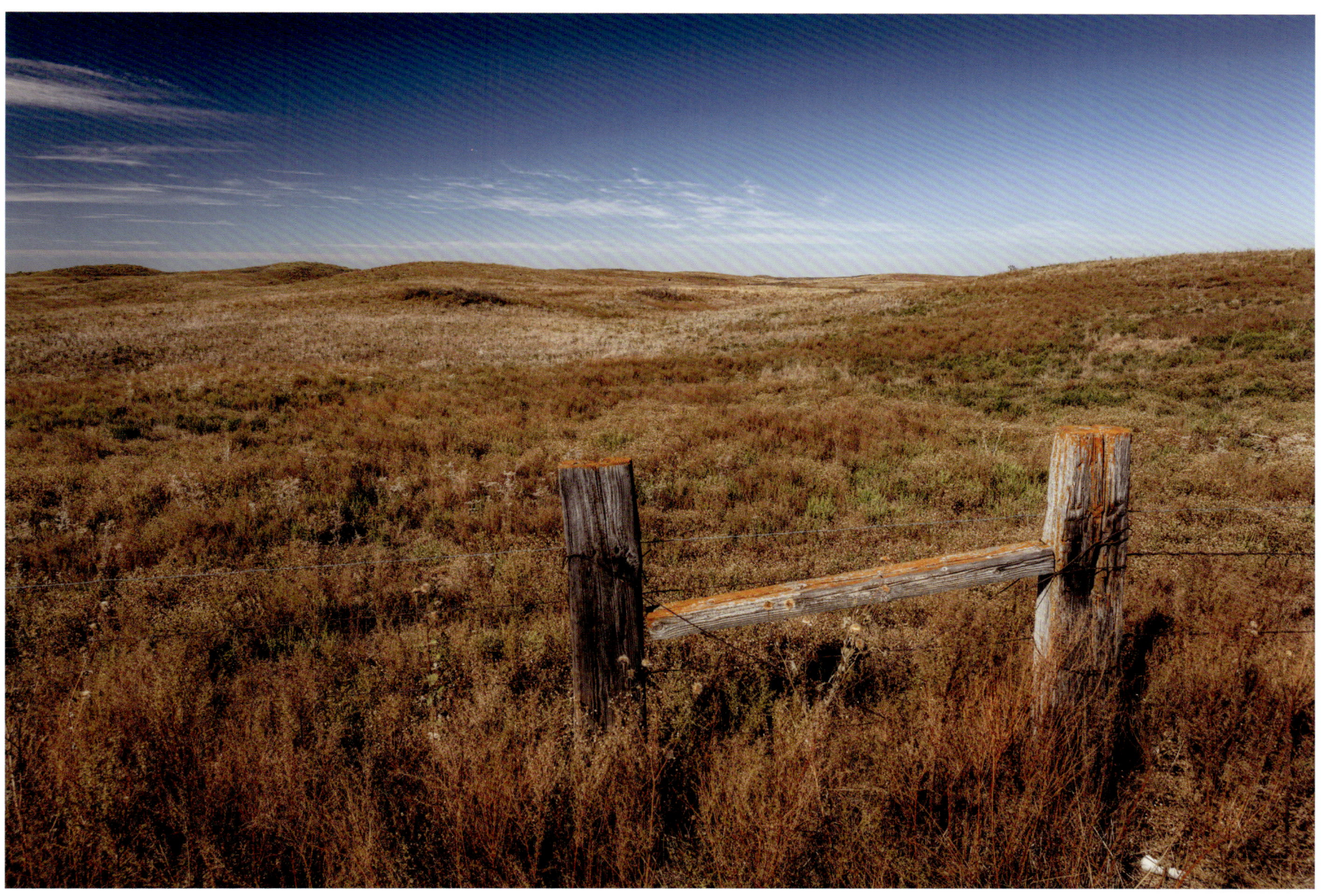

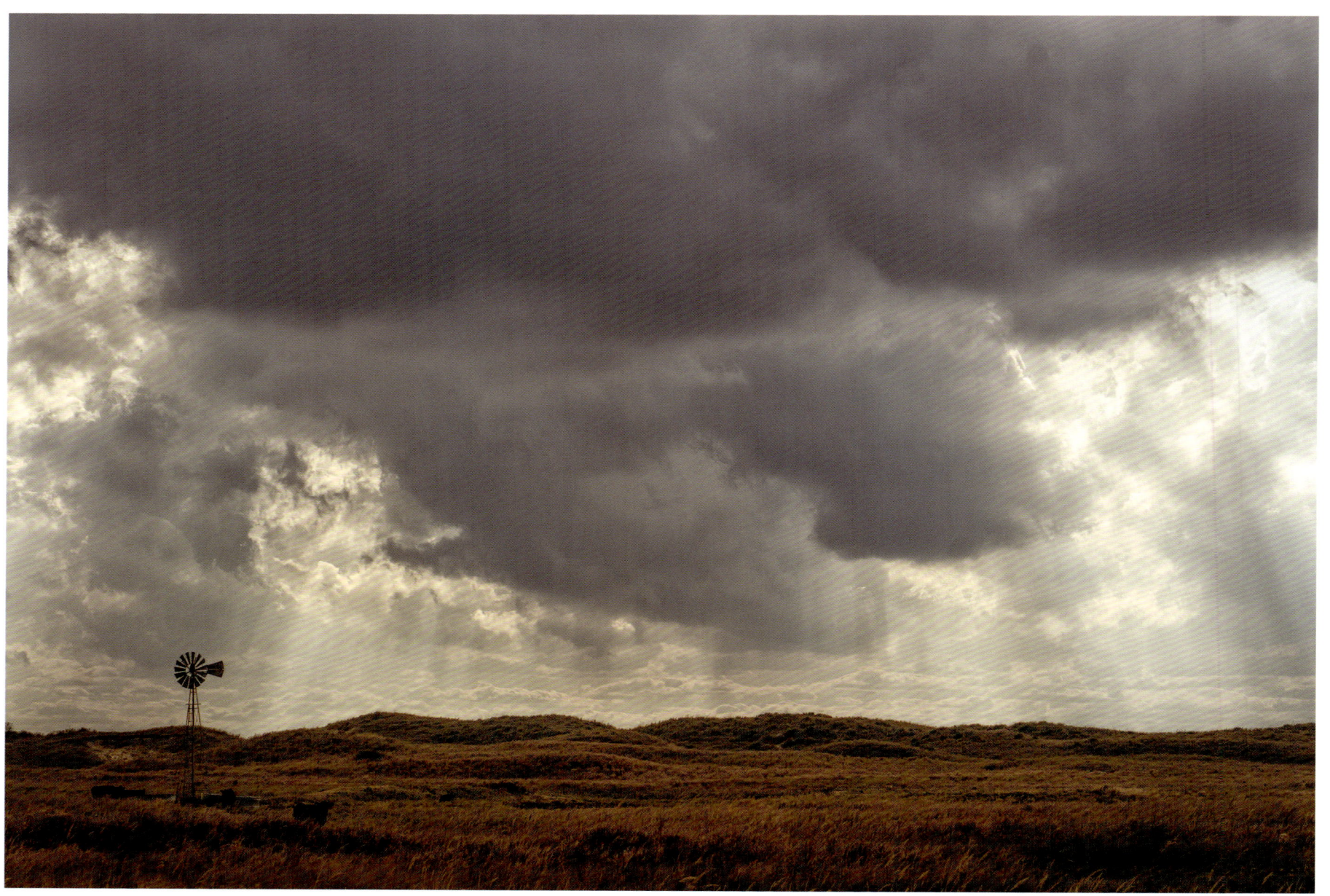

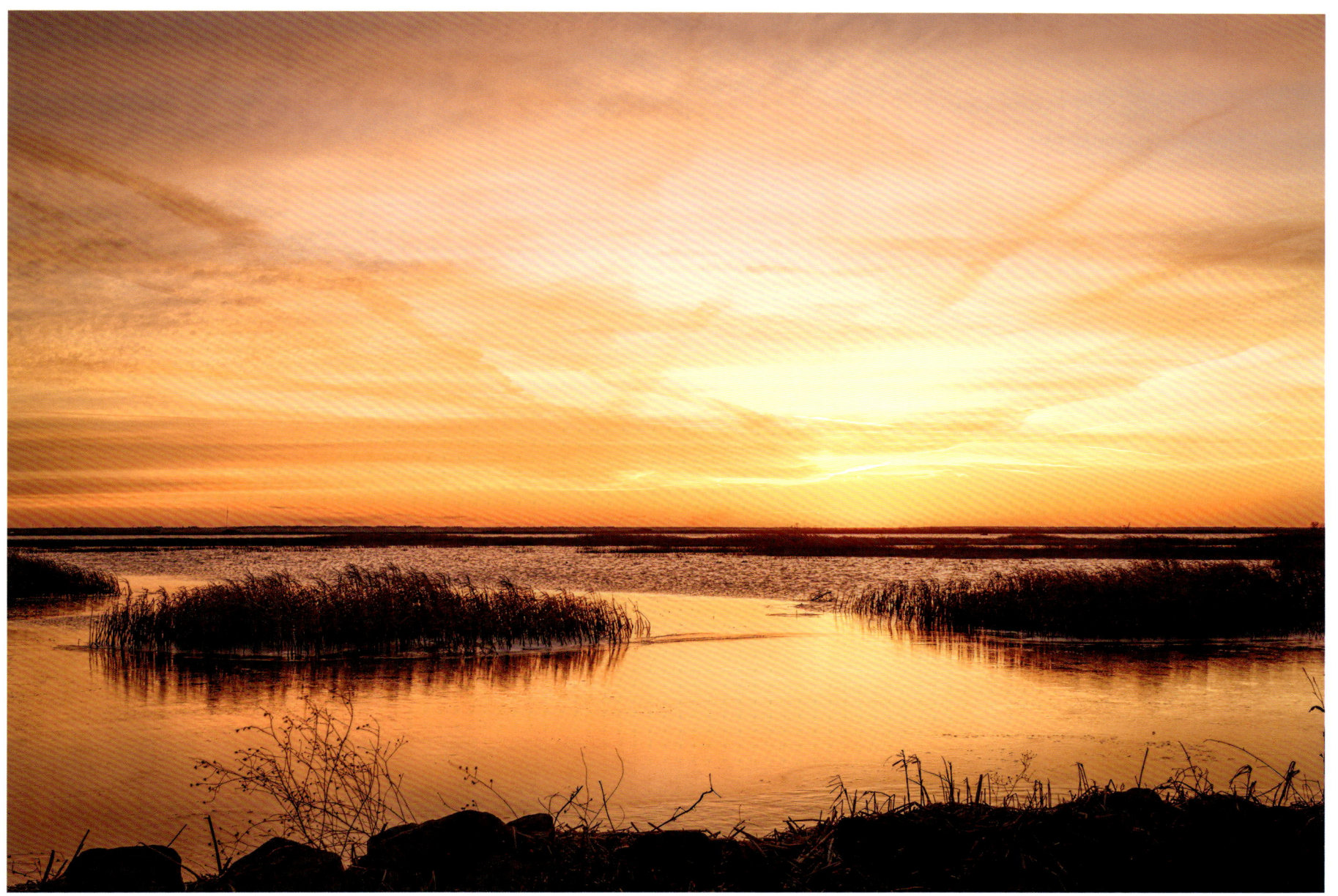

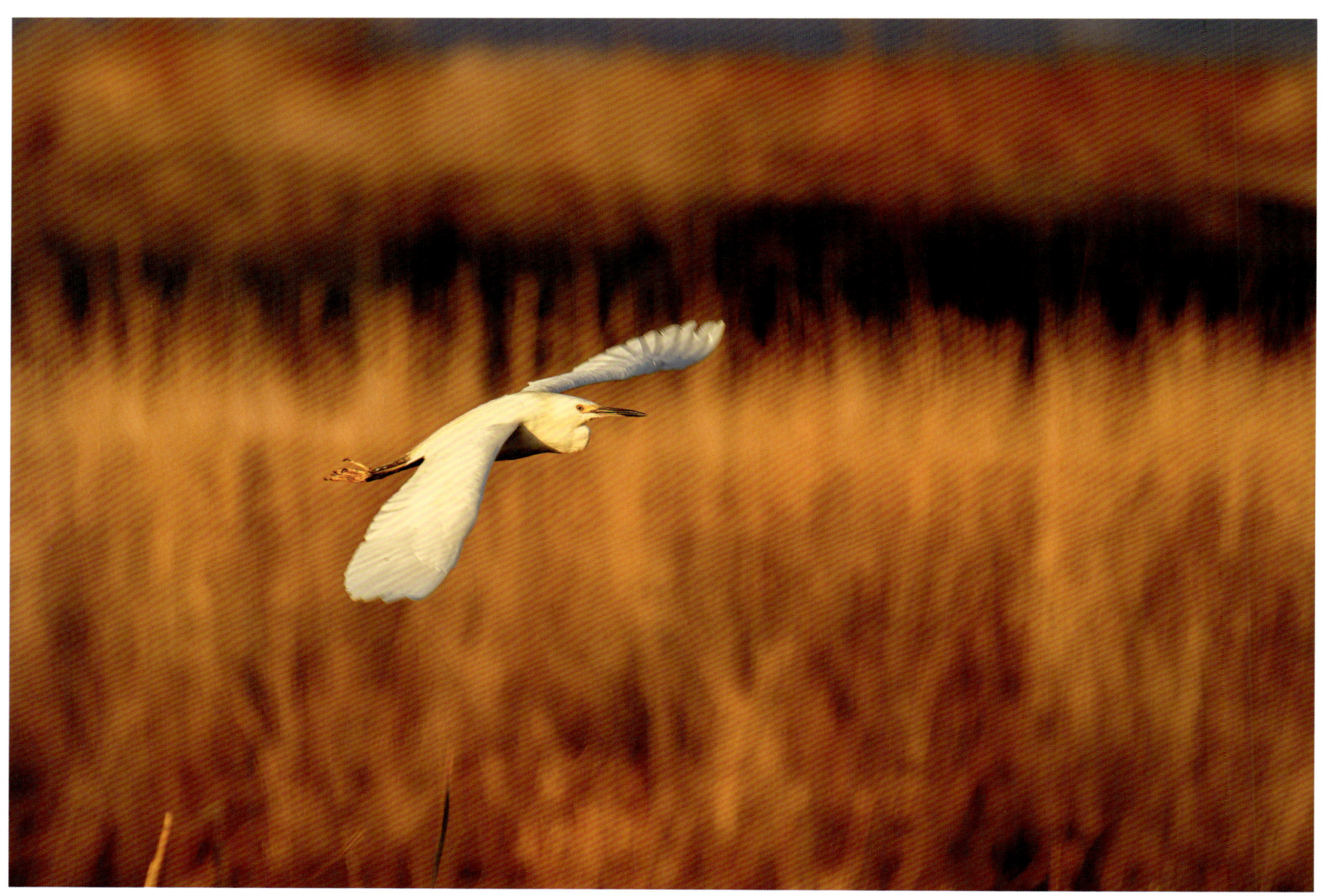

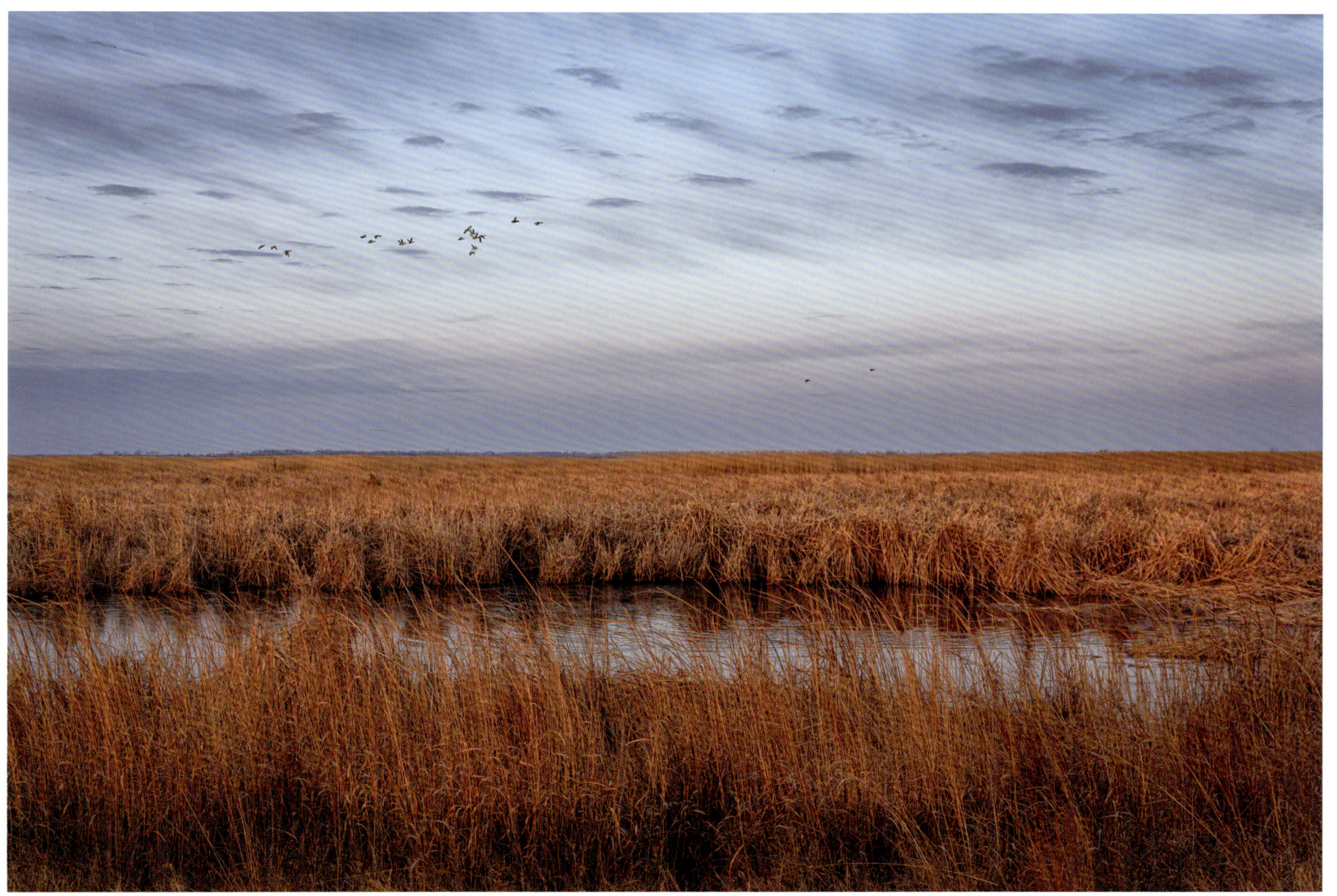

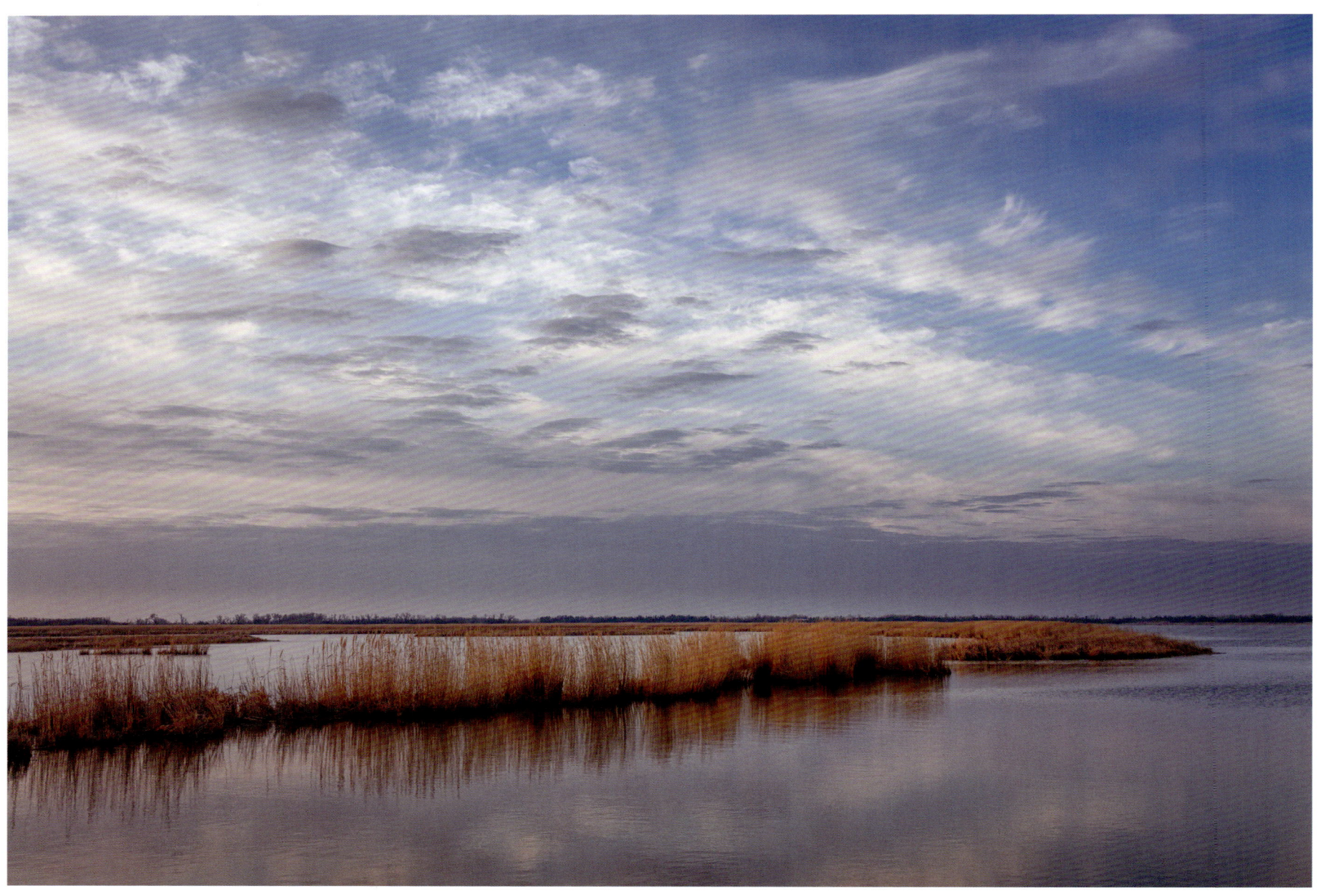

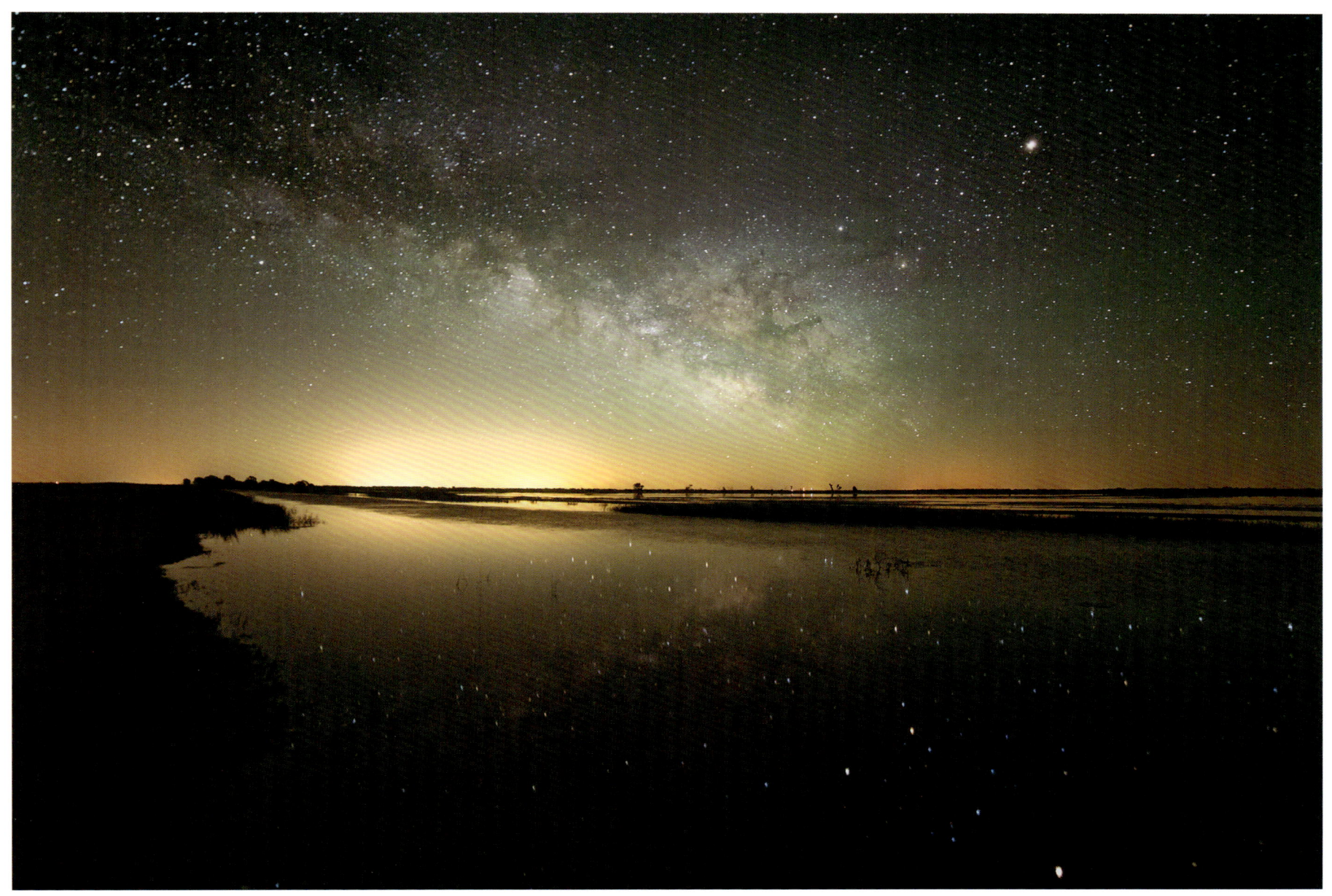

Page 120: I am fascinated by the conditions that cause trees to take on a wide variety of shapes. The size and shape of the trunks, the configuration of the branches, and the height and overall shape create endless patterns for photographs. When I see a tree like this one in Edwards County, I wonder about the specific forces that shaped it.

Page 121: It is interesting to watch the prairie change as you travel west across Kansas. The vegetation in this area of Edwards County has more texture and a different color palette, combining to create a physically harsher feel than what I experience in the eastern prairie regions of the state.

Page 122: The shafts of sunlight coming through the clouds transformed this view of the prairie in Stafford County into a very magical moment.

Page 123: This sunset at Cheyenne Bottoms was a colorful end to a day wandering through the area looking for birds.

Page 124: Many different species of birds, such as this great egret, stop at Cheyenne Bottoms. Driving through the area during peak migration is a special experience, providing the chance to see and photograph a wide range of species.

Page 125: I liked the contrast in the colors of the vegetation and the sky during this late winter morning at Quivira National Wildlife Refuge.

Page 126: The stillness of this morning at Quivira National Wildlife Refuge made for some interesting reflections of the sky in the water of the refuge.

Page 127: A perfectly calm evening at the Big Salt Marsh in Quivira National Wildlife Refuge reflects the Milky Way in one of the small pools below. (Photo by Mickey Shannon.)

Summary

Natural Landscapes: The Results of Geological History, Changing Climates, Evolving Biological Communities, and Human Impacts

In this book, I have sought to communicate the diverse natural processes and vast time spans that underlie the beautiful and subtle landscapes of Kansas. These present landscapes and their ecosystems are a consequence of tens of millions of years of geological processes, changing climates, and biological evolution. Even a brief glimpse of this history gives a deeper appreciation, and even awe, of the familiar natural landscapes that surround us. It is humbling to look back into that deep time. Humanity has been a part of this history for only a tiny portion of the Earth's existence. An awareness of this deep history puts into perspective our relationship to the land. The photographs in this book show Kansas landscapes without the obvious presence of human structures and impacts so that these landscapes can be seen on their own terms. However, the reality is that all of these landscapes have been impacted by human activities.

The midcontinent prairie has been the home of indigenous peoples for thousands of years. In precolonial times, many Native American tribal groups occupied parts of what is now Kansas. These included the Arapaho, Cheyenne, Comanche, Kiowa, Osage, Pawnee, and Wichita. The state is named for the Kansa, or "people of the south wind," who lived along the Kansas and Saline Rivers in central Kansas. These native prairie peoples used only those natural resources that were needed to maintain their communities from year to year and so lived sustainably with their environment. They also manipulated it by fire to draw in the bison and other animals necessary for their food and material needs.

By contrast with indigenous societies, our modern society is built upon unsustainable resource extraction. Our intensive agriculture has stressed available water and reduced the fertility of soils. According to the National Resources Conservation Service, about 190 million tons of Kansas topsoil are lost each year. Streams and rivers have been extensively altered and groundwater supplies depleted. All of our material resources are extracted from the landscape, and much of our energy is drawn from oil, gas, and coal (carbon removed from the atmosphere by plants millions of years ago). The consumption of fossil fuels is impacting the global climate in ways that are stressing and altering both terrestrial and marine biological communities and threatening human populations. Our actions to build our modern societies have also replaced much of the original diverse natural ecosystems with landscapes dominated by low-diversity communities and invasive non-native species. The result is that relatively few areas in Kansas still retain their native biological communities.

In recent years, there has been increasing awareness of the negative impacts of past agricultural practices and resource extraction. Despite our past actions, it is possible to live in a way that respects and preserves the environments that have been bequeathed to us. The more our eyes are open to the beauty, diversity, and underlying history of the landscapes in which we live, the more pressing becomes our desire to find creative ways to live in harmony with those landscapes. I hope that this book can be a small part of raising that awareness.

Acknowledgments

My curiosity about the natural world, and my appreciation for photography, was instilled by my parents, John and Elizabeth Miller. As professional photographers and lovers of nature, they encouraged my growing interest in both the science and beauty of geology. My faculty mentors (Dr. Roger Thomas, Dr. Dick Beerbower, and Dr. Carlton Brett) taught me the skills to reconstruct past Earth environments from the sedimentary and fossil records. Dr. Ronald West at Kansas State University enabled and encouraged my research. I am primarily a field geologist, and my time studying the rock exposures in Kansas gave me a great appreciation both for the diversity of ancient environments preserved in its rock record and for the surprising diversity of its landscapes. When I first met Scott Bean, I was impressed with his ability to capture Kansas landscapes, and this led to our public copresentation on the beauty of those landscapes and the record of Earth history that they disclose. That presentation became the stimulus for this book.

I am indebted to the encouragement and patient guidance of Joyce Harrison at the University Press of Kansas beginning with my first tentative inquiries in early 2022. Her positivity about the project kept me going over all of the hurdles along the way. Kelly Chrisman Jacques professionally shepherded the manuscript through editing, design, and production.

Lastly, I thank my wife, Ruth, who has supported me throughout my geological explorations and was the editor of last resort for this book.

—Keith Miller

Most of my photographs in this book were shot over the past fifteen years or so during my trips around the state. I owe my journey as a photographer to a lot of people. The path that eventually led to me finding myself behind a camera started with family camping trips to Lake Wilson, vacations in Colorado, and fishing and hunting excursions in west and south-central Kansas. I owe my parents a great deal for those opportunities. When I started taking photos, my parents were extremely encouraging; many of the photographs I've taken around Kansas were shot during photo outings with my dad. I wouldn't be where I am today with my photography if it weren't for my parents' support and a great family.

I've also made some good friends through photography, and I'm grateful for their friendship, inspiration, and support. I can't name everyone (and I know I would forget a lot of names), but Ray, Mark, Adam, Michael, Wayne, Eldon, Jim, Brad, and everyone else, thank you.

I've been fortunate to have been helped by many people in getting access to places to photograph and in receiving suggestions about and directions to new places to explore.

In the end, this book wouldn't have happened without Keith taking the lead and making it happen. As I have learned more about the geology of Kansas and the processes that shaped the land, I have begun to look at landscapes in a new way.

I really became serious about photography around the time I married my amazing wife, Jill. She has supported my photography in so many ways and enabled me to do things I would never have been able to do without her strength, encouragement, and wisdom.

—Scott Bean